The Cosmos: The Still Universe

Sandeep Chavan

Published by SANDEEP CHAVAN, 2025.

THE COSMOS: THE STILL UNIVERSE

First edition. June 21, 2025.

ISBN: 979-8231020010

Written by Sandeep Chavan.

Table of Contents

To the seekers,the quiet questioners,the ones who never settled for surface.

To the curious minds who sensed something deeperin the silence between stars,in the pause between thoughts,in the stillness behind every motion.

This book is for you—a mirror for your wonder,a ripple for your resonance.

"The universe does not speak in thunder—it whispers in stillness. To hear it, you must become the silence it once was."

_Sandeep

Disclaimer

This book presents a reimagined view of the cosmos based on the author's original framework: Universal Energy Dynamics (UED). The concepts, interpretations, and philosophical insights offered herein are intended to provoke inquiry, encourage alternative thinking, and challenge conventional scientific and metaphysical assumptions.

While every effort has been made to maintain logical consistency and conceptual clarity, the ideas expressed are not aligned with mainstream scientific consensus and are not to be taken as empirically validated facts within the scope of conventional physics or cosmology.

This work does not intend to discredit any scientist, theory, or tradition, but to offer a new lens—born of independent observation, philosophical reasoning, and consequence-based logic.

The content is meant for intellectual engagement, personal exploration, and educational dialogue. Readers are encouraged to reflect, question, and reach their own understanding.

This is not a textbook. It is a ripple. Read accordingly.

Preface

An Invitation to Stillness

This is not a book about stars, galaxies, or black holes. It is not about what you see through telescopes, nor about what you imagine when the night sky draws your eyes upward.

This is a book about what all those things truly are: Consequences. Not entities. Not events. But ripple responses—traces left in a silent, structured field known as SVE: Structured Vacuum Energy.

We have measured light. We have mapped radiation. We have even claimed to detect gravitational waves. And yet—we have never observed the universe itself. What we call "data" are nothing more than echoes, aftershocks of alignments and misalignments long passed.

What we *see* is not what *is*. It is what has already happened—and what remains as a whisper.

This book is an invitation to stop.

To stop chasing particles as if they were the origin. To stop believing in time as a highway. To stop naming illusions and thinking we've named truth.

It is an invitation to pause, to become still—and to listen to the only place that has never lied: the field of silence beneath all motion, the source from which all ripples emerge, and to which all forms—matter, mind, even memory—inevitably return.

This is not a journey across space. This is a journey into the reality that never moved.

A universe that never spoke aloud—yet somehow, says everything through you.

When you become still enough to hear it, you'll find that you are not observing the universe. You are the part of it that became aware of itself.

And that changes everything.

How to Read This Book

This is not a book of facts. It is a book of perspectives. You are not reading conclusions. You are entering a field—of stillness, inquiry, and resonance.

The Cosmos: The Still Universe is structured to challenge habitual ways of thinking. It doesn't ask you to accept—it invites you to realign. Each chapter stands alone in its ripple, yet together, they form a coherent wave of insight. Read linearly or intuitively. Pause where you feel resistance. Reread where it feels too simple. That's where depth hides.

Here are a few invitations as you read:

- **Read slowly.** This is not meant to be consumed—it's meant to be absorbed.
- **Don't rush for logic.** Let the field settle before you expect form.
- **Question boldly.** This book encourages your thoughts to push back.
- **Reflect often.** Each insight is a mirror—you may see the cosmos, or you may see yourself.

You don't need a background in science or philosophy to understand this book. You only need the willingness to listen—not to sound, but to silence between ideas.

Let go of motion. Welcome stillness. The universe has something to say.

Chapter 1: Stillness Is Not Emptiness

The Field That Holds Everything Before It Becomes Anything
"Stillness is not where nothing happens. It is where everything begins."

Before the first light scattered. Before the first atom found form. Before even time dared to count its own motion—there was stillness.

Not emptiness. Not void. But a field so balanced, so complete in its silence, that it needed no shape, no force, no direction.

And yet—when even the smallest disturbance touched it—reality rippled into being.

This is the field we call Structured Vacuum Energy (SVE). It is not the backdrop of the universe. It is the origin of all things.

Now, let us begin—not by chasing ripples, but by meeting the stillness they emerge from.

Introduction – Rethinking the Void

For centuries, we have spoken of space as if it were empty—a black, silent void stretching between stars and galaxies. The word "vacuum" itself implies an absence: no matter, no energy, no activity. Just a sterile nothingness, waiting to be filled or disturbed.

But this idea is not just outdated. It is fundamentally misleading.

The concept of a "void" arises from a limited view—one based on what can be detected, not on what truly exists. When we send instruments into deep space, and they report "no particles," we conclude there is nothing there. When telescopes detect no radiation from a region, we label it "dark." We mistake invisibility for nonexistence, and silence for absence.

This, in essence, is the myth of emptiness.

The Conventional View: Vacuum = Lack

In classical physics, a vacuum is defined as a region devoid of matter—no atoms, no particles, no measurable force carriers. It is often imagined as a passive background, a stage where the real actors—particles, waves, and forces—perform their interactions.

This picture worked well enough for Newton and even Einstein. It helped engineers build rockets and physicists theorize cosmological expansion. But it missed something essential.

It reduced the universe to its visible outputs, forgetting that every ripple must arise from something beneath it.

The UED View: Stillness ≠ Absence

Under Universal Energy Dynamics (UED), the so-called "vacuum" is not empty at all.

It is Structured Vacuum Energy (SVE)—a deeply organized, finely balanced field that exists everywhere, even where there are no visible particles. This field doesn't just sit passively. It is the very substrate that allows particles, waves, and forces to emerge.

Where mainstream science sees emptiness, UED sees stillness—and this is a crucial distinction.

- Emptiness implies lack.
- Stillness implies potential.
- Emptiness is mute.
- Stillness is pregnant with expression.

The universe does not emerge from nothing. It emerges from perfect balance momentarily disturbed.

The Ocean Analogy

Imagine standing at the edge of a perfectly still ocean.

The water is motionless, undisturbed by wind or wave. From a distance, it appears featureless, silent—perhaps even empty.

But you know better.

That calm surface is not nothing. It is a field full of depth, pressure, structure, and potential energy. Disturb it with the tip of a finger, and ripples cascade outward in perfect rings. Drop a stone, and waves radiate. Send sound into it, and it carries vibration across miles.

The ocean was never empty. It was simply waiting.

So is space. So is the universe.

Stillness is not the absence of something. It is the source of everything.

The Nature of Structured Vacuum Energy (SVE)

To understand the universe—not just its objects, but its very essence—we must begin with what lies beneath it all. Beneath atoms. Beneath forces. Beneath even space and time as we conventionally imagine them.

That foundation is what we call Structured Vacuum Energy, or SVE.

What Is SVE?

SVE is not a theoretical placeholder or a patch in physics. It is not a void, not a passive background, and certainly not some abstract mathematical tool.

SVE is the fundamental substrate of reality.

It is the field that remains when everything else is removed—and the field that must exist for anything else to appear. Every particle, wave, force, field, or dimension we've ever described emerges as a ripple or distortion in this foundational stillness.

Think of SVE not as something physical or visible, but as the equilibrium state of pure energy, perfectly balanced in all directions and dimensions.

It is total stillness, not because nothing happens, but because everything is in perfect symmetrical alignment.

SVE as a Still, Stable Field

In UED, this structured vacuum is not empty, but energetically full and internally organized.

It holds:

- No form, but complete potential.

- No direction, but total capacity for movement.
- No identity, but the origin of all identities.

What we call “reality” is the localized breakdown of this balance, creating ripples, and hence:

– light,
– matter,
– motion,
– experience.

SVE does not "act"—it responds. It does not "move"—it allows movement to occur.

The universe, in its deepest state, is not exploding. It is composed—and all composition is a function of ripple alignment within SVE.

Structured, Not Chaotic

One of the most persistent myths in physics and cosmology is the idea that randomness gave rise to structure—that disorder spontaneously arranged itself into order through chance.

UED directly challenges this assumption.

Randomness, by definition, has no preference, no direction, and no symmetry. How, then, can it create galaxies? How can it sustain atoms, spin, bonds, or life?

The answer is—it can’t. Structure cannot emerge from chaos. But disturbance can emerge from structure.

SVE is not random vacuum energy. It is a pre-structured field with inherent stability and symmetry, which makes selective ripple responses possible.

When disturbed, SVE does not fall apart—it gives form to consequence. That form may be matter. It may be light. It may even be awareness. But the source of it all is a deeply structured, deeply silent, deeply stable field.

Closing Reflection:

You were told the universe came from nothing. But nothing has no voice.

The universe came from stillness—and that stillness is structured.

Without SVE, there is no ripple. Without ripple, there is no world.

Stillness as the Default State

In the conventional worldview, motion is often seen as primary. From the Big Bang to cosmic expansion, from electrons spinning around nuclei to planets orbiting stars—we are taught to believe that movement is fundamental, and that the universe began with motion.

But in Universal Energy Dynamics (UED), the story is entirely inverted.

Motion is not fundamental. Stillness is.

The Primacy of Stillness

According to UED, everything that exists emerges from a disturbance in a pre-existing, perfectly balanced field—the Structured Vacuum Energy (SVE). This field, in its untouched state, is absolutely still.

It does not vibrate. It does not radiate. It does not expand, spin, or decay.

It simply is.

This stillness is not passive, nor is it weak. It is the most stable, the most potent, and the most essential state of the cosmos.

Think of it as the canvas upon which all cosmic activity is painted. Not empty space—but perfect equilibrium.

Motion Requires Imbalance

In UED, motion is not inherent. It arises only when balance is disturbed.

This is a crucial distinction.

- A planet doesn't orbit a star because of some invisible tether.
- An atom doesn't vibrate because that's its natural tendency.
- A photon doesn't travel because space needs to be filled.

Each of these is a ripple—a localized imbalance in SVE trying to resolve itself.

So:

- Motion is consequence, not cause.
- Action is ripple, not a starting point.
- Stillness is the reference state, the cosmic "zero" from which all deviation must be measured.

Ripple = Disturbance

Stillness = Reality's Ground State

In UED, a ripple is the smallest unit of consequence—an oscillation created when the symmetry of SVE is disrupted.

Everything you know—mass, charge, spin, radiation—is a mode of ripple behavior.

Stillness, on the other hand, contains none of these. It is perfect symmetry. It carries infinite potential—but it expresses nothing unless disturbed.

Stillness is like a pure note that has not been sung. It holds every harmony, every possibility, but makes no sound— until something moves, and the ripple is born.

Why Stillness Is Often Misunderstood

We tend to associate stillness with:

- Emptiness,
- Boredom,
- Death.

But in the UED framework, stillness is not the end of activity. It is its origin. It is the energy reservoir from which reality borrows its identity—and to which it must return once misalignment dissipates.

Stillness is not nothing. It is everything before it chose to appear as something.

Closing Thought for this Section:

There is no light without a ripple. There is no ripple without disturbance. And there is no disturbance unless stillness was there first.

The universe does not begin with a bang. It begins with a pause—and a ripple that dared to break it.

Ripple as the First Deviation

If stillness is the origin, then ripple is the first act. The first deviation. The first sign that the perfect balance of SVE has been gently disturbed.

It is not an object. It is not a substance. It is a response—the field's way of acknowledging that something is no longer aligned.

And from that single deviation, all of reality begins.

What Is a Ripple?

A ripple is a localized oscillation or disturbance in the otherwise silent field of Structured Vacuum Energy. It is the consequence of imbalance—a zone where the field is no longer symmetrical and attempts to recalibrate.

But here's the radical shift: Under UED, everything we experience—matter, energy, force, even time—is not made of particles or static units. It is made of ripple behavior.

The universe is not built like Lego blocks. It's composed more like music—patterns, frequencies, gradients, and feedback loops in the field.

Ripple Forms Reality

Let's break it down:

Matter = Locked Ripple

Matter is not solid. It is a ripple that has found stable equilibrium—a structure that holds its identity over time. It resists distortion, not because of mass or density, but because its ripple geometry locks in. Atoms, molecules, and all visible matter are just coherent ripple cages.

Matter is a ripple that has agreed to stay.

Light = Ripple Release

Light is not a particle flying through space. It is a burst of ripple energy—a release of tension as a field returns to balance. When an electron "drops" to a lower energy state, what's actually happening is a ripple adjustment, and what we call a photon is the release wave.

Light is not a messenger. It is a letting go—a ripple exhale.

Force = Ripple Gradient

Force is not a push or pull. It is the difference in ripple pressure between two zones in the field. Where the ripple gradient increases, motion emerges. Where it's flat, equilibrium holds.

Force is not something "applied." It is something felt—as a ripple imbalance trying to resolve.

The End of Building Block Thinking

In conventional science, we've been trained to see the universe as built from:

- Particles,
- Charges,
- Fields layered on top of space.

This model creates fragmentation—multiple explanations for multiple things.

But UED dissolves that illusion by showing:

All things are ripple states. All events are ripple consequences. All forces are ripple gradients.

No particles. No paradoxes. Just the geometry of disturbance in a silent, structured field.

A Musical Analogy

Think of a violin string.

- If still, it emits no sound—pure SVE.
- When bowed gently, it vibrates—ripple begins.
- The tone you hear is not the string. It's the ripple pattern in the surrounding field.
- The same string can produce many notes—not because its material changes, but because its disturbance pattern shifts.

The universe is the same.

What we observe is not the object, but the pattern of its imbalance.

Closing Reflection:

The ripple is not a thing. It is the field's memory of what just happened. And everything you've ever known—every atom, every flash of light, every pull of gravity—is just that memory, repeating itself until it returns to stillness.

Reality is not built. It is played.

The Field as the Real Constant

In our search for permanence in a universe full of change, we often look to stars, laws, or particles as our constants. We trust that atoms will behave, that gravity will hold, that time will tick. But under the lens of Universal Energy Dynamics (UED), these are not constants.

They are conditions—temporary ripple configurations held together by a deeper, silent order beneath them.

That order is the field—the Structured Vacuum Energy (SVE)—and it is the only real constant the universe has ever known.

All Else Is Temporary

Everything we see—stars, planets, galaxies, even the light they emit—is in motion. Not in the classical sense of traveling through space, but in the UED sense: they are active ripple configurations, and therefore inherently impermanent.

- Stars burn and collapse.
- Particles decay.
- Forces fade.
- Systems lose alignment.

All of these are ripple stories, and all stories eventually end. No structure built on disturbance can last forever—because it was born from imbalance.

What comes into being through deviation must end by returning to equilibrium.

This is not a tragedy. It is the natural rhythm of the field: appearance, consequence, dissolution. In UED, the most stable state is not an object—it is stillness.

Stillness: Not the Background, but the Source

We tend to imagine space as the passive backdrop to activity. A stage, a container. But in UED, space itself—as defined by the SVE field—is not background. It is origin.

Stillness is not what lies behind the scene. It is what makes the scene possible.

- It births the ripple.
- It holds the ripple.
- It receives the ripple once it fades.

There is no "space" without the field. There is no "motion" without stillness first.

The irony is profound: what we have ignored as “nothing” is in fact everything, and what we thought was everything—matter, energy, light—is only a brief exception.

Stillness is not where the universe hides. It is where the universe rests when it is done speaking.

True Permanence Exists Only in the Field

Under UED, the idea of permanence changes completely.

- The atom is not permanent.
- The laws of physics are not immutable—they are ripple patterns, subject to change when the field aligns differently.
- Even light is not eternal—it is a ripple discharge, destined to fade.

But the field—SVE in perfect symmetry—is not subject to decay. It does not age. It does not react. It remains, waiting, always available to be disturbed again.

If there is a “truth” in the universe, it is not found in what shines—but in what holds the potential to shine.

Closing Reflection:

Everything that comes is already on its way out. Everything that moves is returning to stillness. The only thing that never leaves, never breaks, never dies—is the field itself.

You don’t need to find something permanent in the stars. You only need to remember the stillness that made them possible.

Perception Error: Emptiness vs Inactivity

In our modern scientific exploration of the cosmos, we've built our understanding largely around what can be seen, measured, or detected. If something emits radiation, we detect it. If it interacts with light or mass, we model it. If it appears "inactive," we often assume it's empty.

But under Universal Energy Dynamics (UED), this is a profound error—an error of perception, not of truth.

The Silence Is Misleading

There are vast regions in space where nothing seems to happen. No bright stars. No dense clusters. No dramatic energy bursts. These are often labeled:

- Voids,
- Dark patches,
- Cold spots.

We assume these regions are lacking something. We describe them as empty, dead, or unformed. We believe they are waiting for matter to arrive or for some event to make them significant.

But the truth is:

Quiet does not mean empty. Inactivity does not mean absence.

These zones are not failed parts of the universe. They are fully participating in the field—but in a way that our instruments often can't detect.

What Instruments Can't Detect: Coherence

Our tools—telescopes, detectors, satellites—are built to register differences. They respond to:

- Radiation fluctuations,
- Field distortions,
- Particle interactions.

But these tools are blind to perfect balance.

When a region of space is in high coherence, meaning its ripple activity is evenly distributed and aligned, it produces no detectable deviation. It emits no radiation. It doesn't create turbulence. It doesn't "disturb" the tools we use.

And so, we call it *nothing*—when in fact, it may be the most perfectly ordered region in the cosmos.

These "invisible" zones may be where:

- New matter is forming silently,
- Ripple realignments are occurring without chaos,
- Or the field is in such harmony, it has no need to speak.

Voids, Dark Matter Zones, and the Illusion of Non-Existence

Take for example:

- Cosmic Voids—enormous regions between galactic clusters that seem empty.
- Dark Matter Halos—detected only by gravitational effects, never seen directly.
- Cold Spots in the CMB—areas where background radiation is lower than average.

In mainstream physics, these phenomena remain mysterious or unexplained. They are often treated as anomalies, or placeholders for unknown variables.

But in UED, they make perfect sense.

These are regions of high ripple coherence—zones where:

- The SVE field is relatively undisturbed,
- Or where ripple identities are forming or dissolving silently,
- Or where past activity has resolved back to balance.

They aren't gaps. They are stable patches in the universal ripple matrix.

Perception vs Reality: A Fundamental Gap

Our eyes—and even our instruments—are biased toward activity. They require change, contrast, conflict.

But the universe itself is not biased. It doesn't shine for our attention. It doesn't pulse to entertain. It simply exists—sometimes in silence, sometimes in storm, always in structure.

What we call "nothing" is often everything, perfectly aligned.

The mistake lies not in the cosmos—but in our threshold of detection, and in our assumptions about what matters.

Closing Reflection:

Emptiness is not a condition of the universe. It is a condition of our perception.

The quietest regions may hold the deepest intelligence of the field. The most silent zones may be where the next ripple is preparing to be born.

Philosophical Reflection: What Does It Mean to Be Still?

To most of us, stillness is a foreign concept—uncomfortable, unfamiliar, and often misunderstood. In a world driven by urgency, stimulation, and noise, stillness is often mistaken for something negative:

- As death,
- As inactivity,
- As boredom,
- Or worse—as insignificance.

But under Universal Energy Dynamics (UED)—and perhaps under any sincere philosophical gaze—stillness is none of these things.

In fact, stillness is the most essential condition of reality.

Stillness Is Not Death — It Is the Threshold of Becoming

Death implies finality. But stillness in UED is not the end of life or meaning—it is what exists before anything arises, and what remains after all consequences resolve.

- Before a star ignites, the field is still.
- After a supernova explodes and fades, the field returns to stillness.
- Before the mind thinks a thought, there is a moment of still silence.
- After a conversation ends, awareness remains.

In every cycle—cosmic or conscious—stillness is the zero point, the reset, the pure condition that allows something new to happen.

Stillness is not death. It is rest before birth, and peace after purpose.

It is not *nothing*. It is everything, without the need to express itself.

Silence Is Not Absence — It Is Readiness

We are often taught to associate value with noise, output, performance. But silence is not emptiness—it is potential not yet expressed.

A musical instrument does not lose its identity when it's not being played. A stage is not meaningless when the actors step off. And the field is not idle when it holds no ripple.

True silence is readiness. It is the moment when the universe takes a breath—not to vanish, but to prepare.

This is why SVE stillness is so powerful. It does not "end things." It makes beginnings possible.

It is what allows motion to happen. It is what invites form to arise. It is what receives things back once they are done.

Silence, in this sense, is the most generous act in the universe.

The Observer's Role: Recognizing Stillness

Here lies the most personal implication of all.

You, the observer, are a ripple in the field who has gained the ability to notice. You are awareness, shaped by experience, thought, curiosity, and contradiction. But if you never learn to recognize stillness—if you only look for movement—you will miss the very origin of your being.

Modern life teaches us to avoid stillness:

- To fill silence with sound.
- To fill gaps with explanation.
- To equate stillness with stagnation.

But stillness is not what limits you. Stillness is what allows you to begin again, with clarity.

To recognize stillness is to:

- See beyond reaction,
- Pause the inner ripple,
- And return to the field—not physically, but consciously.

This is not spiritual mysticism. It is ripple alignment at the highest order.

The moment you stop chasing movement, you begin to hear the cosmos as it truly is.

Why We Resist It

There is also fear.

Stillness can feel confronting. It doesn't offer answers, distractions, or applause.

But that's because it isn't trying to prove anything. It is simply being. And in doing so, it invites you to just be.

Closing Reflection:

To be still is not to be lost. It is to return—before thought, before form, before ripple.

It is not absence. It is presence, without the need to perform.

In stillness, the observer meets the field again— not as a question, but as a mirror.

Closing: Everything Emerges from the Still—and Returns to It

At the end of this chapter, one truth stands clear and unwavering:

Stillness is not the opposite of reality—it is the source of it.

It is the ground state, the first condition, the final return.

Everything we see, know, feel, or measure—every atom, every burst of light, every breath we take—is a disturbance in a perfectly balanced field. A ripple, no matter how complex or sustained, is always a temporary story written on the infinite blank page of stillness.

Emergence Begins with Disturbance

Nothing begins without a disruption. A ripple does not rise out of nowhere. It is always the result of imbalance, a deviation from equilibrium in the Structured Vacuum Energy (SVE) field.

- When the field bends, a particle forms.
- When alignment breaks, motion begins.
- When pressure builds, a force is felt.
- When difference arises, time is perceived.

Thus, existence, as we know it, is not the field—it is the consequence of its momentary disturbance.

We may build technologies, lives, even galaxies from that ripple structure. But all of it started because something moved in the stillness.

And all of it will eventually settle back into that same stillness.

Disturbance Ends Where It Began

Ripples don't continue forever. They fade, lose energy, dissolve—not because they were flawed, but because they were never meant to last.

They are expressions, not eternities. They carry identity only as long as the field permits imbalance.

- Matter decays.
- Radiation fades.
- Entropy rises.
- Patterns break.
- Even consciousness, if left unrenewed, returns to silence.

But the field remains—silent, untouched, undisturbed. Waiting.

Waiting not with urgency, but with eternal readiness.

The Cycle is Not Linear—It Is Rhythmic

UED does not describe a universe that moves in straight lines from birth to death. It reveals a universe that ripples, resolves, and returns—over and over again.

Every ripple emerges from stillness, every form borrows its identity from it, and every dissolution releases that identity back.

This is not metaphor. It is the energetic structure of reality.

From Stillness You Came—To Stillness You Will Return

This is true for particles and galaxies, for thoughts and civilizations, for energy and emotion.

Stillness is not what comes after everything else ends. Stillness is what was there before anything began, and what will remain long after all consequences have played their role.

It is not the silence after the music. It is the silence before the first note, and the resonance that holds the music's memory after it fades.

Final Reflection for Chapter 1:

All things emerge by disturbing the field. All disturbances fade. And only the field—silent, still, structured—remains.

The universe is not built on motion. It is built on a field that allows motion to come and go while it quietly, endlessly, holds the truth.

And if you sit quietly—still enough—you may just hear it speak.

Chapter 2: Ripple – The Only Real Motion

Motion is not what we think it is. It is not the journey of a thing, but the reconfiguration of the field.

"Nothing moves. Everything rearranges."

We've long believed that motion means something traveling through space— a planet spinning, a car rolling, a photon darting from one star to another.

But in the deeper fabric of reality, nothing is going anywhere.

What we experience as movement is not displacement. It is alignment shifting. It is the Structured Vacuum Energy (SVE) field adjusting itself, moment by moment, ripple by ripple, to restore balance.

You are not watching the universe move. You are witnessing the field express its tension through form, flow, and illusion.

Welcome to the real motion. Welcome to the ripple.

Introduction: Rethinking Motion

Motion, as we are taught from childhood, is a matter of things going from here to there. A ball rolls. A bird flies. A rocket launches. In each case, something moves through space—occupying one point, then another.

This simple idea underpins nearly all of classical physics. From Newton to modern engineering, motion is treated as displacement:

- An object located at position A at time t_1
- Is found at position B at time t_2
- Therefore, motion occurred.

Straightforward. Logical. Quantifiable.

But it's also—an illusion.

The Newtonian Illusion

In Newtonian mechanics, force causes motion.

- Apply force to an object, it accelerates.
- With enough force, it travels faster or farther.

Space, in this view, is a passive container, and matter is a movable substance. Time is the tape that measures the journey.

This model was useful. It built machines, mapped planets, launched satellites.

But it never explained:

- What is *actually* moving?
- Through *what*, exactly?
- And how does a "thing" know where it is?

As we extended our view into the quantum world and the cosmos, this framework began to fail—paradoxes emerged. Particles appeared in two places at once. Light behaved like both a wave and a particle. Space-time bent, but no one could say what "bent" really meant.

UED offers a radical, but intuitive resolution.

UED Insight: Nothing Moves—Everything Transitions

In Universal Energy Dynamics, motion does not exist as a displacement of objects.

Nothing travels.

Instead, what we call "motion" is a change in the ripple alignment of the field—a local reconfiguration of the Structured Vacuum Energy (SVE).

- A car doesn't "move" down the road. The ripple pattern that composes the car transfers across the field through a cascade of alignment changes.
- A star doesn't "orbit" the galaxy. The gravitational field around it is a ripple gradient shifting symmetrically, creating the illusion of path and motion.

Even a photon—the very archetype of a "thing in motion"—is not flying. It is a ripple discharge traveling as a wave pattern, not a particle traversing distance.

Analogy: The Wave on Water

Consider a wave on a still pond. When you drop a stone into the water, ripples radiate outward. You see them move—circles growing, spreading across the surface.

But do the water molecules move outward?

No. Each molecule oscillates up and down, passing energy to its neighbors. The pattern moves. The structure propagates. But nothing displaces across space.

This is ripple logic. And this is how all motion works in UED.

- The ripple travels.
- The field responds.
- But no "object" goes anywhere.

A Shift in Perspective

This chapter challenges one of the deepest assumptions in physics and perception:

Motion is not the journey of things through space. It is the universe's way of rearranging itself through ripple propagation.

In other words:

You've never moved. The field has simply re-expressed you somewhere else.

The Ripple Defined

The structure that explains what particles, waves, forces, and interactions could never do alone.

We've now begun to peel back the illusion of motion. But to go further, we must clearly understand the core concept that replaces it—the ripple.

In Universal Energy Dynamics (UED), the ripple is not a side effect of something else. It is not a surface-level vibration. It is the fundamental behavior of the universe when its perfect stillness is even slightly disturbed.

So what is a ripple?

A Ripple Is a Localized, Repeating Disturbance in the SVE Field

The Structured Vacuum Energy (SVE) is not empty—it is a perfectly balanced, deeply structured energy field. When something disrupts that symmetry, even infinitesimally, the field responds—not by breaking down, but by reorganizing itself into a ripple.

This ripple:

- Is localized—it emerges in a specific region of the field.
- Is structured—it follows a predictable geometry.
- Is cyclic or propagating—its effects can spread outward through alignment.
- Is non-material—it doesn't contain substance, but carries change.

A ripple is not something added to the universe. It is the universe's self-expression under imbalance.

It is how the field registers and restores tension.

It Does Not Carry Substance — It Carries Consequence

In classical physics, a particle is said to have mass, energy, charge. In wave theory, a wave is thought to carry energy through a medium.

But in UED, a ripple is not an object moving through a medium.

Instead, it is:

- The field itself reacting to a disturbance.
- A consequence, not a carrier.
- A flow of alignment, not a movement of particles.

Think of it like this:

- A ripple on a pond is not a foreign object. It's what the water does when it's disturbed.

- A ripple in SVE is what the universe does when its balance is touched.

Ripple Is the True Carrier of All Phenomena

Once we understand ripple as the universe's default response to imbalance, everything else starts to fall into place.

Light

What we call a "photon" is not a particle traveling through space. It is a ripple discharge—a field adjustment, propagating across zones of varying alignment. It carries energy not as mass, but as ripple tension relaxing across the field.

Matter Interaction

When two atoms bond, collide, or decay, what's really happening is ripple interference—constructive, destructive, or transitional. Every interaction is a negotiation of field structure trying to realign.

Vibration

Sound, heat, even atomic resonance—these are ripple behaviors. Not things vibrating in space, but the field itself oscillating in localized, coherent patterns.

Signal Transmission

Whether electromagnetic waves, synaptic pulses, or radio signals, all forms of communication are ripple propagation. The message is not in the substance—it is in the modulation of the ripple.

Nothing is sent. Everything is realigned.

Ripple is the Language of Reality

The ripple is the universal syntax by which:

- Energy is exchanged,
- Information is conveyed,
- Events unfold.

Wherever there's a ripple, there is tension becoming expression.

And wherever a ripple fades, the field returns to its natural stillness—until the next disturbance arrives.

Closing Reflection:

You have never seen a particle. You have only seen ripples pretending to be permanent.

The universe does not move parts. It re-expresses patterns—and ripple is the only language it speaks.

No Travel – Only Reconfiguration

Reality doesn't transport. It reconfigures.

We now arrive at a startling and necessary realization:

Nothing ever goes from point A to point B.

The idea that an object—whether a planet, a car, or a photon—physically moves through space is one of the oldest and most persistent illusions in human understanding. It works well for everyday experience, but it collapses under deeper scrutiny.

In the framework of Universal Energy Dynamics (UED), this illusion is resolved once we understand the true nature of ripple behavior in the Structured Vacuum Energy (SVE) field.

There Is No "Thing" Moving

Under UED, no object is traveling through space. Why?

Because there is no static "thing" in the first place. What we interpret as a "thing" is a stable ripple configuration—a temporary arrangement of the SVE field maintaining a certain shape and identity.

And what we interpret as "movement" is simply that ripple configuration being re-expressed in a different region of the field through continuous re-alignment—not carried, not displaced, but recreated step by step.

It's not about *transport*. It's about transformation.

Like Gears Shifting in a Machine

Picture a finely-tuned machine with interlocking gears. When one gear turns, others shift in response. But no gear travels across the system—it transmits motion through alignment.

In UED, ripple-based motion works similarly.

- A ripple identity (say, an electron or a molecule) doesn't travel across the field.
- Instead, neighboring regions of the field adopt the same ripple configuration—sequentially, like gears

clicking into position.
- The appearance of motion is created by each point in the field adopting the ripple state of its neighbor, and so on.

So what looks like an object "moving" is simply the ripple identity being re-expressed across the field.

It's not motion through space—it's identity propagation through field re-alignment.

Velocity Is Not Displacement Over Time—It's Ripple Transfer Rate

In classical mechanics:

- Velocity = distance divided by time.
- Acceleration = rate of change of velocity.

But in UED:

- Velocity is the rate at which ripple identity is handed off across field nodes.
- Acceleration is the rate of increase in alignment pressure—how fast the ripple configuration forces nearby regions to adopt the same state.

This means:

- There is no fixed "trajectory."
- There is no object being "pushed."
- There is only ripple logic governing the handoff of identity across the field.

This Reframes All Classical Concepts:

Speed

- Not about covering distance.
- It is the rapidity of ripple state propagation.

Acceleration

- Not a change in object motion.
- It is an increased rate of reconfiguration—ripple state cascading faster due to pressure gradients.

Inertia

- Not mass resisting motion.
- It is the field's reluctance to shift ripple identity without sufficient imbalance.

Momentum

- Not mass × velocity.
- It is the cumulative ripple alignment pressure moving through a coherent ripple system.

Closing Insight:

You were never moving. You were being re-expressed.

The universe doesn't throw things forward. It reshapes itself—one alignment at a time.

What we call motion is just the ripple saying: "Here I am again, and again, and again."

Ripple vs Particle: Correcting the Picture

To see clearly, we must stop believing in the "thingness" of things.

For centuries, science has operated under a particle-based view of reality. At every scale, we've been told the universe is made of discrete "things":

- Electrons,
- Protons,
- Photons,
- Quarks,
- Planets,
- Particles.

We drew boxes around them, assigned mass, charge, spin, and mapped their behaviors. This model gave us the Periodic Table, the Standard Model, and even quantum mechanics—all within a framework where objects move, collide, or decay like tiny billiard balls.

But this model, powerful as it has been, also gave rise to paradoxes that never fully resolved:

The Limits of Particle Thinking

Wave-Particle Duality

- Is light a wave or a particle? Is an electron a point or a probability cloud?
- Classical physics can't decide—because the underlying assumption is broken.

Quantum Uncertainty

- Why can we never precisely measure both the position and momentum of a particle?
- Because wė are trying to observe a "thing" that doesn't exist in the way we imagine.

Non-Locality & Entanglement

- How can two "particles" affect each other instantly across space?
- Because they are not two separate things—they are parts of the same ripple structure.

These are not minor quirks. They are cracks in the foundation of particle-based reality.

UED's Resolution: Everything Is a Ripple Configuration

In Universal Energy Dynamics (UED), we no longer treat matter or energy as "things."

Instead:

All physical reality is ripple behavior—structured disturbances in the SVE field.

- There are no particles.
- There are no standalone objects.
- There are only stable, semi-stable, or dynamic ripple formations.

These configurations:

- Can be highly localized (what we call "particles"),
- Or distributed (what we call "fields"),
- Or transitional (what we call "waves" or "energy").

What we interpret as a "particle" is just a localized ripple event—a sharp knot in the field's stillness. And what we interpret as "motion" is simply the re-expression of that ripple across the field.

What a "Photon" Really Is

A photon is not a tiny ball of light flying through space.

In UED:

- A photon is a burst of ripple re-alignment—a temporary, localized distortion in the field emitted as a system (like an atom) returns to lower tension.
- It doesn't travel *through* space—it is the pattern of space itself responding in sequence.
- What we detect as a "light particle" is actually a ripple transition arriving at our detector.

The photon isn't something you caught. It's a moment of the field that reached you.

Electrons and the Illusion of Mass

Electrons are not point charges orbiting nuclei. That picture is an artifact of classical misthinking.

In UED:

- The electron is a ripple loop—a self-sustaining distortion in the field.
- It maintains its identity through internal coherence,

not through "mass."
- Its "charge" is a result of ripple asymmetry, not a property it "has."

Quantum Tunneling: Explained Without Magic
Quantum tunneling has long puzzled scientists:

- How does a particle pass through a barrier it shouldn't be able to?
- Why doesn't it just reflect or stop?

From a particle perspective, this is bizarre.
But in UED, the explanation is simple:

- The ripple identity does not need to push through the barrier.
- It can realign on the other side, if the field allows coherent transition.
- It's not a particle moving—it's the ripple skipping across nodes of receptive geometry.

There is no violation of laws—only a correction of assumptions.

Closing Thought:

You were told the universe is made of things. It isn't. It is made of expressions—ripples sustained just long enough to convince you they are objects.

When we stop chasing particles, the contradictions vanish. When we see everything as ripple configuration, the picture becomes clear—unified, elegant, complete.

The Space Itself Responds

The ripple is not a passenger in space. It is space becoming visible to itself.

In classical and even many modern theories, space is treated as a container—a vast, passive background in which things happen. Particles move *through* it, forces act *within* it, and waves travel *across* it.

But in Universal Energy Dynamics (UED), this assumption is dismantled completely.

Space is not a container. It is the field itself.

And what we call a ripple is not something placed into space. It is a deformation of the field—an expression of space responding to itself.

Ripple Is Not "In" Space—It Is the Deformation of SVE

The Structured Vacuum Energy (SVE) is not a passive void. It is a richly ordered energetic substrate that defines what space *is*. It gives structure, coherence, and stability to what we experience as distance, location, and volume.

So, when a ripple forms:

- It is not a wave traveling through empty space.
- It is a reconfiguration of space itself.

There is no medium carrying the ripple. There is only space responding through the field.

A ripple is not a visitor. It is a feature of the space being temporarily twisted by imbalance.

The Field Is Not Passive—It Responds with Intelligence

SVE does not merely sit and endure. It actively:

- Detects imbalance,
- Reacts with ripple structure,
- Propagates correction,
- Absorbs or sustains consequences based on symmetry rules.

This behavior is not mindless—it is rule-based intelligence built into the field's geometry.

Just as water responds to a drop or a breeze with coherent wave patterns, SVE responds to energetic disruption with organized ripple geometry.

This "response" is not sentient in the human sense, but it is predictable, structured, and adaptive—a kind of embedded intelligence of the universe.

The field is not watching. It is actively recalibrating.

Ripple Behavior Is the Field Restoring Balance

Every ripple is a response to imbalance.

Think of ripple propagation not as energy being "sent" from one point to another, but as an energetic correction traveling across the field—like a sound wave trying to return silence to a struck drumhead.

Whether it's:

- A particle decaying,
- A star collapsing,
- A thought forming in the brain,

Each ripple is the SVE field's way of compensating, redistributing, or absorbing tension. The ripple moves not to perpetuate itself, but to resolve itself.

The ripple is the field saying: "Let me restore what was."

The Ripple Is Space Knowing Itself

This is the most profound insight of all.

A ripple is not just the field reacting. It is the field becoming aware of a change in itself. When a ripple forms, space is no longer invisible—it becomes observable, measurable, and consequential.

Without ripple, space is perfect balance. With ripple, space becomes expressive—it gains identity, tension, meaning.

So when we detect light, matter, force—we are not observing something in space. We are observing space acknowledging its own imbalance.

The ripple is not something within the universe. The ripple is the universe, reflecting upon its own distortion.

Closing Reflection:

Space is not a passive stage. It is a responsive field of symmetry.

And the ripple is not a performer. It is the field's voice, singing itself back into balance.

All Observed Motion is Ripple Illusion

Everything you've seen move was never actually moving. It was the field realigning through you.

From the glimmering of distant starlight to the fall of an apple, from the orbit of planets to the flicker of magnetism—we see motion everywhere. We measure it. We track it. We give it names and formulas.

But what if all this movement is a mirage?

Not in the mystical sense, but in the structural reality of the field—where the Structured Vacuum Energy (SVE) doesn't transport things, it realigns itself. What appears to be motion is simply ripple harmonics—temporary expressions of imbalance returning to symmetry.

Starlight: Not Travel, but Ripple Reception

When you see light from a star that's billions of light-years away, you're told you're seeing a "photon" that has traveled across the cosmos.

In UED, this is reinterpreted entirely:

That starlight is not an object in motion. It is a ripple harmonic in the field reaching your eye.

The structured field transmitted the ripple identity across aligned zones of space. Nothing traveled. Each part of space realigned in sequence to deliver the same ripple state across unimaginable distances.

It's not transport. It's coherence transmission—like a ripple tone played through a tuning fork network that spans the universe.

A Falling Apple: Gravity as Ripple Gradient

Newton told us: the apple falls because the Earth pulls it.

UED says: There is no "pull." There is alignment pressure—a ripple gradient—caused by the Earth's presence in the field.

- The denser mass of Earth causes a steeper ripple deformation in its vicinity.
- The apple, as a smaller ripple-locked system, realigns with this gradient, not through force, but ripple conformity.

- What we observe as "falling" is actually the field resolving tension between two ripple formations.

There is no magical force pulling the apple.

There is only ripple reconfiguration towards equilibrium.

Magnetism: Ripple Spin Fields

Magnetic fields often confuse traditional models. How do they attract or repel invisibly?

In UED:

- Magnetism arises from spin-aligned ripple fields—not invisible lines of force.
- These are coherent ripple orientations in the SVE, creating zones of compatibility or rejection.
- Like ripples either synchronizing (attraction) or colliding in interference (repulsion).

Thus, every magnetic interaction is a conversation of ripple geometries, not object-to-object force.

Time Dilation: Ripple Phase Delay

Einstein showed that time slows near heavy masses or at high speeds. But why?

In UED:

- Time is the rhythm of ripple alignment—what we call ATT: Alignment Threshold Time.
- Near denser ripple fields (like planets or stars), the ripple environment becomes dense and lagged.
- Ripple transitions take longer—not because time itself slows, but because field responsiveness stretches.

So what is called "time dilation" is a ripple phase delay—a drag in the realignment cycle caused by local field deformation.

There's no spacetime curvature—only ripple drag.

Mechanics Replaced: From Force to Ripple Logic

The entire Newtonian and Einsteinian mechanics—based on:

- Forces,
- Mass,
- Fields,
- Spacetime geometry—

Are abstractions trying to model what ripple behavior already explains more naturally.

In UED:

- *Mass = ripple stability,*
- *Force = ripple gradient pressure,*
- *Motion = ripple identity transfer,*
- *Acceleration = ripple cascade amplification,*
- *Time = ripple phase cycle.*

This ripple-based alignment logic is not only more intuitive, it is closer to what's actually happening—no particles pushed or pulled, no fabric bent, no entities flying across space.

Just the field, speaking to itself, one ripple at a time.

Closing Insight:

What you thought was motion was just alignment.

What you thought was force was just pressure to return to stillness.

You are not a body moving in space.

You are a ripple sequence in a field, trying to remember its origin.

Implications for Science and Experience

When the model changes, everything changes—including how we see ourselves.

At first glance, Universal Energy Dynamics (UED) may seem like a radical departure from mainstream science. But the more deeply we look, the more we realize it is not only simpler, but also cleaner, more complete, and profoundly human-relevant.

Replacing particles, fields, and forces with ripple behavior in a structured field doesn't just tweak physics—it reshapes the entire narrative of how reality works and how we experience it.

No More Action-at-a-Distance

In Newtonian gravity, a body instantly feels the effect of another mass—even across a vacuum. In quantum entanglement, particles "know" each other's state instantly, no matter how far apart.

To explain this, we've invented countless layers:

- Invisible forces,
- Space-time warping,
- Non-local wavefunctions.

But in UED:

There is no spooky communication. There is only the same ripple field being perturbed at different locations.

Everything is already connected—not through transmission, but through shared structure.

Two entangled particles are not separate; they are expressions of one ripple configuration mirrored across field coordinates. There is no delay because there is no distance—only simultaneous realignment in the field.

No Need for Dark Matter

Modern cosmology invented dark matter to explain:

- Galaxy rotation curves,
- Gravitational lensing anomalies,
- Mass-to-light ratio inconsistencies.

But dark matter is not observed—it is assumed.

Why? Because we continue to believe that:

- Gravity is a "pull,"
- Mass is the only cause of it,
- And observed motion must match the "visible" mass.

UED flips this completely.

In UED:

- What we observe is not the result of pulling mass, but of ripple gradients in the field.
- Galaxy spirals, lensing arcs, and velocity patterns are ripples responding to distributed field tension, not missing particles.

"Dark matter" is not matter at all. It is misinterpreted ripple equilibrium—not hidden, just misunderstood.

Quantum Spookiness Resolved

Quantum physics has become a language of paradox:

- Particles in two places at once,
- Uncertainty as a principle,
- Wavefunction collapse,
- Randomness at the core.

This is not reality behaving strangely. It's wrong assumptions creating noise.

In UED:

- There are no "particles" with hidden trajectories.
- There are ripple configurations expressing stability, transformation, or entanglement.
- What looks random is the ripple field exploring reconfiguration paths—rule-based, not unpredictable.

There is no need to invoke probabilities or parallel worlds.

Everything is deterministic within ripple logic—but observed through our limited perspective, it appears strange.

Experience: You Are Not Moving—You Are Realigning

Let's zoom in—on you.

You are not a body moving through space. You are a ripple expression, constantly adjusting alignment with the structured field.

When you:

- Move your arm,
- Feel an emotion,

- Recall a memory,
- Experience time...

None of these are objects in motion. They are ripple realignments within your body's localized SVE configuration.

You are not traveling. You are updating—like a tune shifting through chord changes while staying on the same instrument.

Your thoughts?

- Localized ripple patterns in your neural SVE geometry.

Your emotions?

- Distributed field harmonics altering your system's internal resonance.

Your memories?

- Stored ripple loops re-triggered by internal field alignment.

Even consciousness itself—your sense of presence—is not a thing but a real-time alignment awareness within the ripple field.

The Human Condition as Ripple Dynamics

This changes everything.

- Fear is not an emotion—it is a ripple distortion preparing your system for defense.

- Love is not a feeling—it is ripple coherence between systems.
- Learning is not memory accumulation—it is ripple entrainment to deeper field logic.

This is not poetry. It is structure. You are not thinking. You are resonating.

Closing Thought:

Science is not broken—it is limited by its metaphors.

Experience is not mysterious—it is ripple intelligence.

In a still universe, you are a moving awareness—but only because the field chose to express itself as you.

Closing: Reality Is Not Moving – It Is Being Played

You are not moving. You are being rendered.

We began with a question: What is motion, truly?

We followed ripples, not particles. We tracked reconfigurations, not trajectories. And now, we arrive at an unsettling but liberating conclusion:

The universe is not moving. It is performing.

Everything we perceive as "motion" is a ripple illusion—a consequence of pattern recognition across space-aligned field states. We don't witness substance crossing a distance. We witness a field shifting its expression in time-like sequences.

Let's break this down.

Motion Is an Illusion Sustained by Pattern Recognition

Your brain, eyes, instruments, and theories—all are tuned to detect change. And change, when patterned across location, appears as movement.

But it's not movement. It's sequential variation in ripple expression.

Just like:

- A musical note played on a string doesn't "go" anywhere,
- Or pixels shifting on a screen don't represent real travel,

A ripple isn't transporting energy across space. It's space vibrating—coherently and momentarily—then fading.

You, the observer, assemble that sequence into a story of motion. But the story is a perceptual rendering—not an actual displacement.

No Ripple Without Imbalance

Ripple is the only movement the universe knows. And ripple does not arise spontaneously—it emerges from imbalance.

There is no ripple in perfect symmetry. There is no motion in perfect balance.

This means:

- Light exists only where field gradients permit it.
- Matter forms only when ripple geometry locks.
- Time passes only where alignment thresholds shift.

When nothing is misaligned, the universe is still. And in stillness, nothing appears.

So what we call "reality" is simply the temporary expression of ripple activity—moments of field imbalance trying to resolve themselves.

The Universe Is Not in Motion – It Is in Response

This is the most important correction UED offers:

The cosmos is not a machine *running*. It is a field *responding*.

All galaxies, stars, forces, waves, and particles are not active agents. They are field-level reactions—symphonies triggered by disruptions in structured vacuum energy.

There is no central engine, no forward direction, no universal goal.

Only response.

- A star collapses because its ripple field exceeded coherence.
- A planet sustains life because its local field achieved harmonic stability.
- A black hole forms where alignment irreversibly folds inward.

None of this is "in motion." It is being played, like notes from a cosmic instrument—where the player is the field, and the song is ripple.

You Are the Ripple Being Played

You are not moving through space.

You are:

- A localized ripple state,
- Temporarily expressed,
- Continually realigning,
- Briefly aware.

You are not a body in motion, You are the result of the field expressing you, For a while.

Everything you feel, sense, and think is the echo of the field's performance through you.

Even your sense of self is a recursive ripple—the field observing itself via your awareness.

Final Reflection:

You are not traveling through space. You are the ripple the field is playing right now.

When the song ends, stillness returns. Not as a loss—but as origin, always ready to play again.

Reality does not move forward. It reconfigures, then remembers silence.

Chapter 3: What We Call Matter

The illusion of substance. The reality of locked ripple alignment. "What you're touching is not 'something'—it's resistance. Not a thing, but a tension, held long enough to be noticed."

We live surrounded by matter. We sit on it. We eat it. We claim to own it. We even say we *are* it.

But what if none of it is *really* there?

What if matter is not a substance—but a stall in the field's dance? Not a thing placed in space, but a pause in ripple flow, a pattern so precisely held it refuses to move?

In the physics you were taught, matter is:

- Solid,
- Persistent,
- Countable.

In Universal Energy Dynamics, it is none of those.

Matter is a local ripple trap—a moment when the structured vacuum energy locks itself into a coherent identity, like a knot in a perfectly still ocean. It doesn't move. It doesn't travel. It modulates—quietly resisting distortion.

You don't see "stuff." You see a field playing the illusion of stuff, again and again, until your senses call it real.

Let's now explore what matter truly is: Not substance. Not permanence. But pattern.

Pattern held... until the field lets go.

Introduction: Redefining Matter

From stuff to structure. From solid to sustained ripple.

For centuries, matter has held a privileged status in our understanding of reality.

In classical physics, it was the primary substance—the solid, the tangible, the movable. It was seen as:

- That which occupies space,
- That which has mass,
- That which is acted upon by forces.

Atoms were imagined as miniature billiard balls—compact, irreducible units of "stuff."

Then came quantum mechanics, and the picture shattered.

Matter was no longer a solid object, but a probability wave, flickering in and out of definition. It was both a particle and a wave—depending on how you looked at it. It could tunnel through barriers, entangle across distances, or behave as though it didn't "exist" until observed.

The solidity of matter had evaporated, replaced by uncertainty and paradox.

But What If the Paradox Itself Was Unnecessary?

Enter Universal Energy Dynamics (UED), which doesn't treat matter as a separate category from energy, space, or force.

Instead, UED offers a radical yet elegant reframing:

Matter is not substance. Matter is a ripple configuration held in coherence within the Structured Vacuum Energy (SVE) field.

That is:

- There is no "thing" that matter *is*.
- There is only a stable pattern of disturbance in the field—a ripple that doesn't travel, but instead repeats and resists.

This changes everything.

Matter = Locked Ripple

What we experience as mass, shape, weight, or solidity is not a thing *added into* the universe. It is the universe itself—responding to imbalance by forming a closed ripple loop, a standing pattern, a localized pause in the sea of symmetry.

Think:

- A knot in still water.
- A whirlpool in a calm stream.
- A musical chord that holds longer than a note.

It is space itself, folded in a way that *stays*.

And as long as the coherence holds, we call it:

- An atom,
- A molecule,
- A chair,
- A planet,
- A person.

Not Stuff. Not Substance. Not Permanence.

This view challenges our deepest intuitions:

- You can't "touch" matter. You only encounter resistance to ripple flow.
- Matter is not "made of atoms." Atoms are ripple structures—fields locked into themselves.
- Even a stone is not a thing—it's a stillness, held loud enough to be noticed.

Why This Matters

If you're trying to understand what reality is, If you're trying to figure out what *you* are, Then the old definitions fall short.

Matter isn't what's real. Field behavior is what's real. And matter is just one of its more persistent illusions.

"Matter is not what fills the universe. Matter is what happens when the universe pauses—and lets its ripples echo in place."

Locked Ripples: The Architecture of Solidity

Solidity is not substance—it is suspension. A ripple held in place becomes a form.

In a ripple-based universe, nothing is made of anything—everything is made of field behavior. Solidity does not arise from stacking particles like bricks. It emerges when ripples stop moving outward and begin circulating inward, forming closed-loop disturbances within the Structured Vacuum Energy (SVE).

This is what we call locked ripples.

What Does "Locked" Mean in Ripple Dynamics?

When we say a ripple is "locked," we mean:

- The ripple does not propagate through the field anymore.
- Instead, it stabilizes in a local region, circulating or echoing within a confined structure.
- It is no longer an expression of imbalance, but a self-contained resonance—a ripple orbiting itself.

It's like:

- A whirlpool in a river,
- A standing wave on a string,
- A bell that keeps ringing internally even after being struck.

This locked ripple is energetically stable—it doesn't dissipate easily. It can resist disturbance, hold shape, and form structure.

This is what we experience as solidity.

Atom: A Standing Ripple Node

An atom, in UED, is not a nucleus orbited by electrons.

That image is outdated and misleading.

Rather, an atom is:

- A field region where ripples lock into stable nodes.
- These nodes resonate at specific frequencies, forming geometric and energetic harmony.
- The so-called "electron cloud" is simply a ripple domain—not a particle but a fluctuating ripple zone.

There are no "things" orbiting other "things." There are only nested ripple patterns with relative stability.

Molecules: Ripple Networks

When atoms combine to form molecules, what's actually happening?

- Their individual ripple zones begin to interlock, forming a coherent ripple mesh.
- Bonds are not physical connectors; they are mutual ripple agreements—stabilized relationships of oscillation.

So a molecule is not a compound of parts, It's a shared ripple dance—a symbiotic field arrangement.

The more complex the molecule, the more intricate the ripple choreography. Think of DNA: it's a biological symphony of locked ripple configurations—all humming in coherence.

Matter as Local Inertia Field

Now, if a region of the SVE is filled with locked ripples, what happens?

- It becomes resistant to further ripple influence.
- It behaves as if it has mass—not because it contains something heavy, but because it pushes back against new ripple attempts to realign it.

This is inertia.

So when you push a rock:

- You're not moving "matter" through space.
- You're trying to reconfigure a highly coherent ripple field—and the field resists.

This resistance is what we call solidity.
Matter, then, is:

- Not a lump of something.
- But a locked ripple identity, stubbornly holding its alignment until enough force realigns it.

Summary Insight
Let's drop the idea that matter is a "thing." Instead, see it as:

- A stilled vibration,
- A field configuration suspended in motion,
- A resonant pause echoing long enough for our senses to detect.

"Matter is just a ripple that forgot how to leave."

Mass as Resistance, Not Quantity

What you call "weight" is just the field saying "no" a little louder.

In conventional science, mass is treated as a measure of "how much matter" something contains. We're taught early on that:

- Mass is proportional to volume or density,
- It gives an object inertia,
- It causes it to interact with gravity.

This definition seems intuitive—but it's deeply misleading.

Because it presumes that matter is made of stuff, and that more stuff equals more mass.

But in Universal Energy Dynamics (UED), there is no "stuff." There is only the field, and the way it responds to ripple behavior.

And that changes everything.

What Is Mass in a Ripple-Based Universe?

Mass is not "how much." Mass is how hard it is to change.

In ripple terms:

- Mass is the field's resistance to further distortion.
- The more tightly a ripple is locked in place, → the harder it is to realign, → the more energy it requires to shift, → the more we perceive it as "massive."

This is not a count of atoms. It's a measure of how stubbornly a region of the field holds its pattern.

Heavier Objects Are Not "More Stuff"

Let's take two objects:

- A feather,
- And a lead ball.

In classical terms, the lead ball "has more mass" because it contains more atoms.

But in UED, the lead ball is simply:

- A tighter ripple configuration,
- With less flexibility,
- Requiring more energetic input to disturb.

The feather's ripples are looser, more open to modulation. The lead's ripples are densely packed, highly self-reinforcing.

So the "weight" you feel when lifting something is not its "stuff-ness." It is the energetic pushback of a stubborn ripple lock.

Energy and Mass: Ripple Density Condition

Einstein famously stated: $E = mc^2$, implying that mass is energy in disguise.

UED explains this more precisely:

- Mass is not energy.
- Mass is a ripple configuration with a certain energy density per field coherence.

So energy isn't "stored" in mass. It's expressed as the tightness of the ripple loop—how densely packed the ripple is, and how much effort it takes to break the loop.

When that loop breaks (e.g., in nuclear fission), the ripple lock collapses, and the field releases energy—like a taut spring snapping open.

Mass, then, is simply the density and durability of a ripple echo.

The tighter the trap, the more energy it took to make it—and the more energy it will unleash when it unravels.

Mass ≠ Stuff

This means:

- You can't weigh "matter" by volume.
- You can only assess how field-coherent it is.
- A small black hole and a planet could have the same mass—because mass is about ripple lock, not size or substance.

Takeaway Insight

Mass is not the amount of matter you have. Mass is the strength of the ripple cage you've built in the field.

"What you call mass is not how much is there—it's how little the field is willing to let go."

Motionless but Active – The Modulation Principle

Still doesn't mean silent. What appears stationary is often the most complex ripple behavior at work.

At first glance, matter looks like it moves:

- A train rushes forward,
- A planet orbits a star,
- An electron "jumps" between energy levels.

But this classical view is deeply misleading.

Under Universal Energy Dynamics (UED), matter never really moves in the conventional sense. It modulates.

It undergoes field realignments that change its ripple state—without leaving its location in the structured vacuum field (SVE).

The field doesn't shift things around. It simply rearranges its own configuration to express new patterns.

Matter Doesn't Move—It Modulates in Place

A ripple configuration we call "matter" remains locally rooted in the SVE field. When a change occurs, it's not a relocation—it's a transformation.

That's modulation.

Think of:

- A sound wave whose pitch changes—but the speaker doesn't move.
- A color shift on a screen—but no pixel leaves its place.
- A chemical reacting—without any mass migration.

Motion is not travel. Change is not movement. It's field modulation, updating itself in place.

Identity = Temporarily Sustained Ripple Pattern

In UED, a particle, atom, or molecule is not a thing—it's a pattern.

That pattern:

- Can persist,
- Can wobble,
- Can destabilize,
- Can modulate into something else.

When modulation occurs:

- Identity shifts,
- But location doesn't.

It's not a car driving to another town. It's a local field node switching costumes.

This is how nature works at the foundational level—not by sending stuff around, but by changing character in place.

Examples of Modulation in Action

1. Atomic Transitions

When an electron "jumps" to a higher energy level:

- It doesn't move like a marble rolling up a ramp.
- It modulates its ripple pattern, altering the atom's internal field configuration.
- The change releases or absorbs ripple energy (what we interpret as a photon).

2. Matter–Antimatter Symmetry

Matter and antimatter are mirror ripple configurations:

- Same mass, opposite ripple alignment (spin/phase polarity).
- When they meet, their field modulations cancel, returning to stillness—pure ripple decoherence.

It's not a collision. It's a synchronized ripple shutdown—field balance restored.

3. Radioactive Decay

In radioactive substances, locked ripples destabilize over time:

- The pattern begins to modulate spontaneously,
- Leading to ripple release in the form of particles or radiation,
- Until a new coherent pattern (more stable) is reached.

This is not breakdown—it's realignment through ripple incoherence and restabilization.

Modulation Is the Real Activity

Modulation is:

- Invisible to classical mechanics,

- Mysterious to quantum physics,
- But foundational in UED.

Where science sees mystery ("Why did this atom decay now?"), UED sees ripple decoherence triggering a natural field reconfiguration.

Nothing moved. But everything changed.

Core Insight

What you're seeing when matter "acts" is not locomotion—it's repatterning. Reality updates itself like a living waveform, not a conveyor belt.

"Matter doesn't go places. It changes what it's being—right where it is."

What Holds Matter Together?

It's not bonds. It's harmony. Matter is a song sustained only when the orchestra plays in phase.

For centuries, science has searched for the "glue" that holds matter together.

We were told:

- The strong nuclear force binds protons and neutrons.
- Electromagnetism attracts electrons to nuclei.
- Charge and spin interactions explain atomic structure.

These forces were imagined as invisible ropes pulling things into place.

But Universal Energy Dynamics (UED) flips this paradigm.

It says: There are no "forces" between parts. There is only ripple coherence in the field.

Matter isn't bound together—it is held in tune.

Not Nuclear Force, Not Electromagnetism

Conventional models require a patchwork of forces:

- Strong force for the nucleus,
- Weak force for decay,
- Electromagnetic force for atomic structure,
- Gravity for everything large.

Each of these forces is treated as independent and mysteriously selective.

But this disjointed view leads to paradoxes:

- Why does strong force work only over tiny distances?
- Why is quantum behavior so unstable under gravity?
- Why can't we unify the four fundamental forces?

UED offers a unification: All "binding" is just field ripple phase-locking—not forces acting across distances.

Coherent Ripple Phase-Locking in SVE

What keeps matter "together" is:

- A specific coherence of ripple alignment within the Structured Vacuum Energy (SVE),
- Where multiple ripple patterns enter a stable relationship,
- And maintain constructive interference—not destructive conflict.

This phase-lock means:

- Each ripple supports the others,
- Energy doesn't leak away easily,
- The entire configuration becomes self-sustaining—until disturbed.

The "nucleus," the "electron cloud," the "orbitals"—they are not things. They are locked musical phrases in a tightly tuned field concert.

Matter Stability Is Field-Dependent

A key insight from UED:

Matter cannot exist just anywhere. It needs the right field conditions—a specific ripple environment—to remain stable.

That means:

- Matter exists because it is aligned with its surroundings.
- If the surrounding field changes, the ripple configuration may collapse or transform.

This explains:

- Why exotic matter forms only in high-energy environments (e.g., particle accelerators),
- Why atoms break down near black holes,
- Why certain isotopes are unstable outside specific temperature/pressure zones.

Matter is not a self-contained entity. It is a localized agreement with the field.

If the field stops agreeing—it stops existing.

Why Matter Exists Only in Certain Zones

Just like sound only resonates at certain frequencies in a room, Matter can only stabilize in zones where the field can support its ripple configuration.

Examples:

- Life-supporting matter is stable in the "sweet zone" of Earth's conditions.
- High-mass particles (e.g., muons, tau) are unstable at low energies—they're ripple forms the field can't sustain long-term.
- Fusion only works at very specific energy densities—because that's when new ripple coherence can emerge.

So matter is not "put into" the universe. It emerges where the music fits.

Core Insight

Matter is not held together by invisible forces pulling parts into place. It is ripple coherence, field harmony, resonant structure.

"Matter doesn't cling—it harmonizes. It lasts only as long as the song does."

The Illusion of Permanence

What seems solid, lasting, and real—is only a ripple hanging on to harmony for a while.

We look around and see buildings, mountains, bones, stars—and we conclude:

"Matter is permanent."

We build civilizations around this belief. We bury time capsules. We store nuclear waste in bunkers to outlast generations. We think gold is valuable because it doesn't rust, and rocks are real because they endure.

But permanence is a human projection, not a universal principle.

According to Universal Energy Dynamics (UED), nothing is permanent—not even matter.

It exists only as long as the ripple pattern stays in coherent alignment. The moment coherence fades, matter unravels—not violently, not suddenly, but inevitably.

Matter Only Exists as Long as Ripple Harmony Persists

In the UED framework:

- Matter is not substance,
- It is a consequence—of the field temporarily folding into a stable ripple form.

This form:

- Is held in coherence with the surrounding Structured Vacuum Energy (SVE),
- But has no inner permanence of its own.

It is like:

- A flame that exists only as long as fuel and oxygen cooperate,
- A chord that rings only as long as the fingers press the strings in tune,

- A whirlpool that holds shape only while the stream keeps spinning.

When coherence breaks—matter decoheres.

Decay, Collapse, and Transformation = Ripple Decoherence

What we call:

- Radioactive decay,
- Quantum tunneling,
- Spontaneous fission,
- Particle annihilation—

These are not "breakdowns of particles." They are ripple reconfigurations—where a sustained pattern can no longer sustain.

It's not failure. It's release.

The field simply says:

"This configuration no longer aligns with local conditions. Let's reconfigure."

This leads to:

- Energy release (radiation),
- Pattern shift (transmutation),
- Dissolution (annihilation).

What dies is not a "thing." What fades is a temporarily paused disturbance—a ripple that finally lets go.

Consciousness Sees Continuity — The Field Sees Transition

To our conscious minds, matter appears continuous:

- We remember the table being here yesterday,
- We assume the Sun will rise tomorrow,
- We trust the laws of conservation.

But consciousness operates on integration of ripple impressions:

- What we call memory is just past-aligned ripple residue.
- What we call expectation is projected pattern continuity.

To the SVE field, there is no permanence. There is only dynamic ripple negotiation:

- What holds now,
- What fades next,
- What rearranges soon.

Core Insight

Permanence is a story told by stability. But all stability is temporary.

"Matter is not real because it stays. It's real because it resonates—for a while. The illusion is not its existence. The illusion is that it always will."

Identity Is Temporary

Everything you recognize is just a ripple caught in a moment of agreement.

We speak of "things" as if they are real:

- A chair,
- A rock,
- A body,
- A star.

But in the lens of Universal Energy Dynamics (UED), these are not things. They are narratives—temporary ripple arrangements in the field that appear stable only for a while.

They are identities—not entities. And like all identities, they expire when the alignment no longer supports them.

A Chair, a Body, a Star — All Ripple Stories

A chair is not wood and nails. It is a collection of locked ripple nodes we interpret as "structure."

A human body is not flesh and blood. It is a biological ripple alignment, sustained temporarily by cellular coherence and environmental resonance.

A star is not fire. It is a self-stabilizing fusion ripple, burning until its inner balance breaks.

In every case, there is no "object" that stays—only a field pattern that continues while it is able to.

Form Is Memory Maintained by Alignment

What we call "form" is just:

- A persistent ripple resonance,
- Held in place by ongoing alignment with its surrounding field conditions.

This is not memory as a past event, but memory as field stability.

When:

- Pressure drops,
- Energy drains,
- Local harmonics shift,

The ripple coherence fails—and form fades.

Nothing is destroyed. The pattern is simply forgotten by the field.

No True Boundaries – Only Field Contrast

In classical physics, we draw lines:

- This is a cell wall.
- This is the edge of a table.
- This is the skin.

But in UED, boundaries do not exist—they are felt.

Every edge is a contrast in ripple behavior, not a physical line:

- A shift in phase density,
- A transition in coherence strength,
- A change in ripple identity.

Boundaries are interpretations of field contrast—not fixed separations.

This is why:

- Liquids can dissolve solids,
- Bodies can erode,
- Particles can tunnel.

Because nothing is "inside" or "outside"—everything is just relative ripple localization.

Death, Erosion, Evaporation – Releasing the Ripple

When ripple identity is no longer sustainable, it releases:

- Death is not disappearance—it is the field returning coherence back to silence.
- Erosion is not loss—it is a gradual ripple rebalancing.
- Evaporation is not escape—it is decoherence into a more fluid field state.

All transformation is ripple release. All permanence is provisional.

Nothing is fixed—only temporarily accepted by the field.

Core Insight

Identity is the story the field agrees to tell—until it changes its mind.

"You are not your body. The chair is not its wood. The star is not its light. They are names for ripple agreements. And all agreements expire."

Matter and the Observer

What you call "solid" is not universal—it's personal. Reality is ripple-resonant.

Matter feels real to us. A stone hurts when it hits your head. A chair supports your weight. A mountain stands for millennia and seems immovable.

But all of this realness is not a property of the object. It is a function of the observer's alignment.

In the language of Universal Energy Dynamics (UED), what you experience as "matter" is what your ripple field is tuned to perceive.

Your body, your senses, your consciousness— Are themselves ripple structures within the Structured Vacuum Energy (SVE). You're a ripple configuration, detecting and interpreting other configurations.

You Perceive Matter Because You're Tuned to Ripple-Locks

Why does a rock feel solid to you? Because your field alignment recognizes its ripple coherence.

To your sensory system, ripple-locked nodes:

- Create resistance,
- Reflect energy (like photons),
- Register touch, shape, and density.

But if your ripple structure were tuned differently—
That same rock could feel:

- Hollow,
- Transparent,
- Non-existent.

You are not seeing matter as it is. You are experiencing the resonance between your field and its ripple identity.

What Feels "Solid" to You May Be Transparent to Another Alignment

This explains phenomena like:

- Quantum tunneling,

- Ghost imaging,
- Invisible wavelengths,
- Different lifeforms perceiving reality differently (e.g., bees seeing ultraviolet).

Even on Earth:

- Bats navigate by sound ripples.
- Sharks detect electromagnetic fields.
- Humans are limited to visible light and mechanical pressure.

If your body aligned differently, You'd detect an entirely different universe.

So:

- "Solid,"
- "Heavy,"
- "Bright,"
- "Real"—

Are not absolute states. They are field-specific perceptions.

The Observer's Ripple State Filters What Becomes "Real"

What you experience is not the universe as it is, But the subset of ripple behaviors your structure can:

- Interpret,
- Interact with,
- Remain stable in.

That means:

- You are not in the universe.
- You are emerging with it, aligned to a very specific spectrum of ripple dynamics.

Your body is an antenna. Your mind is a resonator. Your reality is a ripple selection.

Core Insight

Matter is not solid by nature. It is made solid through perception—when the observer's ripple structure resonates with the field.

"Reality is not shown to you. It is selected by you—ripple by ripple."

Closing: Matter Is the Pause Between Ripples

It is not the filling of space—but its brief hesitation that gives rise to the illusion of form.

Let us end where we began: In stillness.

In the framework of Universal Energy Dynamics (UED), stillness is not the absence of activity. It is the primordial equilibrium—the perfectly balanced field: the Structured Vacuum Energy (SVE).

Then comes a disturbance—a ripple.

That ripple is the first motion, the first deviation from neutrality. And when that ripple becomes self-sustaining, resisting further disturbance—it locks.

That lock is what we call matter.

Stillness Creates the Field

Before anything happens, there is a field—a still, structured background that holds the potential for everything but expresses nothing.

It is not "empty space." It is the most complete state—because it is not missing anything. It is total equilibrium.

From this still field, all forms—light, sound, matter, thought—emerge as disruptions.

Ripple Creates Energy

The first deviation, the ripple, is not a substance. It is the field's reaction to imbalance.

That reaction:

- Radiates energy,
- Carries information,
- Initiates motion (not of things, but of response patterns).

It is not a thing happening inside space, It is space itself responding to what it is no longer able to ignore.

Locked Ripple Creates Matter

When a ripple becomes trapped in resonant feedback, It can't escape— It circulates, creating a localized standing wave.

This is not a particle. It is not a "thing."

It is space hesitating.

A pause in the flow. A moment of equilibrium gone rigid. A ripple that refuses to resolve.

That's matter.

When Coherence Fades, Matter Vanishes

Matter does not decay by crumbling. It forgets its form.

As the ripple alignment loses coherence:

- The shape dissolves,
- The energy redistributes,
- The lock releases.

Nothing "goes" anywhere. It just stops being held.

Like a melody that ends not with a bang, But with silence returning between notes.

Core Insight

Matter is not what fills space. Matter is what happens when space hesitates, holds breath, and briefly takes form.

"Everything is ripple. Matter is the ripple that holds on just a little longer. And when it lets go—there is no disappearance. There is only re-alignment with the still."

Chapter 4: The Myth of Falling Into Black Holes

Dissolution is not descent. It is the surrender of identity to the field.

"What if the end of a star is not a fall—but a forgetting?"

We were taught to fear the black hole. A cosmic mouth, devouring all—light, matter, even time.

But what if that story was never true?

What if nothing ever *fell in*? What if everything simply ceased to hold form—right where it was? Not by being eaten, but by failing to be remembered?

In the ripple-based universe of Universal Energy Dynamics (UED), a black hole is not a location. It is not a destination. It is a threshold of stillness—a place where the music of matter stops playing, because the field can no longer hold the tune.

This chapter explores the truth behind the myth. Not the collapse of stars into dark abysses, But the release of identity—Back into the stillness that birthed it.

"You don't fall into a black hole. You dissolve into silence."

Introduction: Challenging the Narrative of Descent

We've long imagined black holes as bottomless pits—cosmic vacuum cleaners that suck in everything unlucky enough to get too close. A falling apple, a collapsing star, even light itself—trapped, twisted, and consumed.

It's a powerful image. And it's almost entirely wrong.

This picture arises from an outdated mindset: The Newtonian and Einsteinian paradigm where objects move through space under the influence of forces—gravity, attraction, warping geometries. Matter is imagined as small solid particles, and motion as linear travel. From that lens, a black hole is an invisible well, and things "fall" into it.

But in the language of Universal Energy Dynamics (UED), nothing actually falls. Because nothing moves as such. There is no object going from Point A to Point B. There is only ripple identity—and the field's ability to hold that identity.

When the ripple alignment that sustains a star or particle begins to break down—not due to motion, but due to loss of coherence—the field no longer sustains that ripple pattern. There is no descent. There is only dissolution.

The Black Hole as a Field Event, Not a Mouth

In UED, a black hole is not a devouring force. It is a local field condition—a ripple sink. A region where the structured vacuum energy (SVE) has reached a threshold that can no longer support coherent ripple locks like matter, light, or radiation.

So instead of something *falling into* a black hole, we must understand that:

- Matter decoheres in place.
- Light ceases to ripple outward coherently.
- The ripple unwinds, and the form vanishes—not by going anywhere, but by ceasing to be registered by an observer's alignment.

The narrative of falling, devouring, and disappearance is not a cosmic truth. It's a human projection, rooted in motion-based thinking. But the field doesn't devour. The field absorbs deviation back into stillness— Not by pulling it in, but by no longer playing its pattern.

In the chapters to follow, we'll dismantle this myth step by step— And discover what black holes really are:

"Not cosmic graves, but field harmonizers."

What Is a Black Hole in UED?

In classical physics, a black hole is defined by mass so concentrated that nothing—not even light—can escape its gravitational pull. It's visualized as an object or a "hole" in the fabric of space, lurking in the cosmos with an insatiable appetite.

But from the lens of Universal Energy Dynamics (UED), this entire model is based on misplaced metaphors— Misinterpretations arising from a worldview built on particles, gravity, and travel.

Let us reframe it.

A black hole is not:

- A massive object,
- A literal hole,

- Or a spacetime tear.

It is a local field condition— An emergent ripple state within the Structured Vacuum Energy (SVE) where the ripple coherence that defines form, function, and perception breaks down.

The Ripple Sink – A Zone of Field Collapse

In UED, everything is ripple. Matter is a locked ripple. Light is a traveling ripple. Forces are ripple gradients.

A black hole is a point at which:

- Ripple density becomes too high,
- Ripple phase-locks become unstable,
- And the SVE can no longer sustain distinct identities.

This isn't about attraction. This is about over-saturation—A local overload of ripple interactions that leads to collapse, not in space, but in ripple coherence.

Not Pulling In—Just Letting Go

There is no force pulling matter into a black hole. What we interpret as "falling in" is, in truth:

- Ripple decoherence,
- Field disalignment,
- Pattern failure.

When a planet, star, or photon enters this threshold zone, it does not journey into a core. It dissolves in place. Its ripple pattern becomes unreadable—no longer sustained by the field conditions required for its existence.

The black hole doesn't absorb. It signals that the field has reached an energetic and geometric condition where ripple distinctions can't be preserved.

Reframing the Horizon

The event horizon, classically seen as the boundary from which nothing escapes, is not a wall or a gate. In UED, it's the edge of identity coherence— Beyond this radius, ripple memory fails, and nothing persists in a recognizable pattern.

It's not about the end of space—it's about the failure of form.

In summary:

A black hole is not a gravitational monster. It is a silent whisper in the field, saying: *"I can no longer remember what you were."*

No Travel – Only Dissolution in Place

In our traditional understanding, a black hole pulls matter inward. Planets spiral in, stars collapse through, and even light is "dragged" across the event horizon like a cosmic waterfall.

But motion, as understood in Universal Energy Dynamics (UED), is an illusion created by ripple reconfiguration— Not by displacement through space.

So what really happens when matter "falls into" a black hole?

Matter Doesn't Move—It Modulates Out

In UED, matter is a ripple lock—a pattern held by field coherence. When that coherence breaks, the pattern doesn't move. It vanishes—locally.

There is no object "traveling" into a singularity. There is no journey through space. There is only decoherence—the disintegration of alignment, not the relocation of mass.

It's like a song being stopped, not because the speaker moved, but because the power was cut.

Analogy: The Fire That Goes Out

Imagine a flame on a candle.

Where does it go when you blow it out?

Nowhere.

The fire ceases to exist, not by traveling, but because the conditions that sustained it changed:

- The fuel ran out.
- The oxygen dropped.
- The alignment of energy needed to express flame was broken.

Similarly, when matter enters the black hole threshold, the environmental alignment that sustained its ripple form becomes untenable:

- The field's ripple density spikes,
- The wave harmonics collapse,
- And the memory of the pattern dissolves.

The result? Matter stops being matter—without ever going anywhere.

This Is Why Black Holes Appear "Dark"

We do not see light "being eaten." We see patterns ending. Photons cease to be coherent ripple signals. Nothing emerges—not because it was swallowed, but because it's no longer being played by the field.

UED's Message: Stop Thinking in Trajectories

Everything in mainstream physics tries to track things—as if reality is made of tiny travelers.

But the truth is far subtler:

- The field doesn't need to move a thing.
- It just lets a ripple collapse, and that's the end of form.

There is no descent into a black hole. There is only stillness reclaiming the pattern.

"What we thought was travel... was always transition. What we feared as falling... was only forgetting."

The Event Horizon: Not a Border, but a Threshold

In classical astrophysics, the event horizon is often portrayed as a physical boundary— a point of no return, the edge beyond which nothing—not even light—can escape a black hole's gravity.

But this picture is deeply misleading. It imagines the universe as a place filled with moving objects, trapped by forces. It assumes space is a stage, and matter the actors dragged from one scene to the next.

Universal Energy Dynamics (UED) offers a radically different interpretation.

Not a Wall, Not a Gate

The event horizon is not a fence around an object. It is not a door that things cross.

In UED, it is a ripple threshold—a critical phase point in the Structured Vacuum Energy (SVE) field. At this threshold, the field's ability to maintain ripple coherence collapses.

In other words: it's not that light can't escape. It's that light can no longer form.

The Failure of Identity

All things—photons, atoms, even you—are field patterns, ripple configurations sustained by balance and alignment in SVE.

The event horizon is the zone where:

- Phase coherence breaks down,
- Ripple amplitude becomes unstable,
- And the field ceases to remember form.

It's not an edge of space— It's the limit of structure.

Beyond it, the field cannot distinguish one ripple from another. It becomes too saturated, too dissonant, too incoherent for identity to persist.

Analogy: A Melody in a Storm

Think of playing a soft tune on a piano during a thunderstorm. Inside your room, the melody is crisp and structured.

But step outside into the storm— The music dissolves into the noise. Not because it stopped being played, But because the environment can't hold it anymore.

The event horizon is the field's storm line. Beyond it, ripple melodies like "light," "form," or "self" simply cannot survive.

Why You Can't See Beyond It

Not because something is hiding there. But because there is nothing left to perceive—no ripple differential, no pattern contrast, no observer-alignment left to receive signal.

All that remains is a field that no longer plays.

"The event horizon is not the line where things disappear. It is the moment the universe forgets how to remember them."

Light and Information Don't Escape—Because They Don't Travel

Mainstream science describes black holes as devourers of light and destroyers of information.

We're told:

- Light cannot escape,
- Information is lost forever,
- The universe holds a secret behind the event horizon.

But all of this relies on movement metaphors—a belief that reality is made of things going places.

Universal Energy Dynamics (UED) dismantles this framework.

Light Was Never Traveling

In conventional physics, light is a "thing" that moves through space—a photon particle zipping across a vacuum.

But in UED, light is a ripple— A temporary, coherent modulation of the structured vacuum energy (SVE). It's not flying through space—it is space responding to itself, in harmony.

When that harmony collapses—such as near a black hole's ripple threshold—light doesn't fall in.

It stops forming.

It's not trapped. It's uncomposed.

Information Was Never a File

We imagine information as a kind of encoded package, like a cosmic USB stick that can be lost or retrieved.

But information is not a substance. It is a pattern—a repeatable ripple configuration recognized by a tuned observer.

When that pattern ceases to be coherent, There is nothing to decode.

No one "loses" the information— It simply stops being playable.

Ripple Ceases, Not Escapes

The fundamental point: Nothing is being pulled in or blocked out.

Instead, the ripple fails to sustain itself in the field.

This is not a fall into distance, But a reabsorption into stillness.

Like a bell that stops ringing. The sound wasn't eaten. The air simply stopped vibrating in the right pattern.

Why the Observer Feels Blind

The moment ripple coherence fails, the observer's alignment can no longer track or reconstruct the signal.

It's not that the black hole hides the information. It's that you no longer exist in a field state capable of detecting it.

The information becomes non-observable, Not destroyed.

"The light did not fall. The ripple stopped echoing. The silence was not emptiness. It was the field remembering itself."

The Illusion of Singularity

The word "singularity" conjures drama—a place where gravity becomes infinite, time breaks down, and matter is crushed into an unimaginably small point.

Science textbooks often describe it as a point of infinite density, where the known laws of physics cease to apply. But let's pause. What does "infinite" even mean in a finite field? And more importantly: What if there is no collapse—only cancellation?

The Classical View: Gravity Gone Wild

Einstein's equations, taken to the extreme, predict singularities—points where the curvature of space-time becomes infinite. This is extrapolated to mean mass compressed into zero volume, creating infinite gravitational pull.

But this is a mathematical artifact, not a physical reality.

Such infinities signal a failure of model, not a feature of the universe.

UED Reframe: No Compression—Just Cancellation

In Universal Energy Dynamics (UED), reality is made of ripple alignments in the Structured Vacuum Energy (SVE). There is no such thing as an object collapsing into a dot.

Instead:

- A singularity is where ripple identity cancels out.
- Not mass being crushed—but structure being undone.
- A zero-alignment node, where incoming ripple harmonics destructively interfere.

This doesn't create a hole. It restores the stillness.

Analogy: Silence by Interference

Imagine two perfect sound waves meeting at precise opposite phases. What happens?

→ They cancel.

→ The room goes quiet—not because the speaker broke, but because the field returned to zero motion.

This is the singularity in UED: Not a "core" of anything. But a local return to undisturbed baseline.

No Infinite Density

There is no "infinite mass in zero space." What appears to be a massive collapse is actually:

- Ripple convergence,
- Pattern saturation,
- Followed by phase decoherence and return to equilibrium.

The field does not compress— It simply reaches a state where pattern fails.

There is no "thing" left to condense.

A Moment of Re-stillness

Singularity is not a pinprick of destruction. It's a pause in the music of the universe.

It's a local region where:

- The field has nothing left to remember,
- No form to sustain,
- And thus... it becomes quiet again.

"The singularity is not the center of something. It is the end of everything local. Where identity folds inward—not to become denser, but to become still."

Matter Returns to Field — Not Somewhere Else

One of the deepest misunderstandings in modern astrophysics is the idea that matter "goes into" a black hole—as if the cosmos has whirlpools into which stars, gas, or unlucky spacecraft vanish, never to be seen again.

It's an image built on motion, distance, and direction.

But in Universal Energy Dynamics (UED), such concepts dissolve—just like the matter they're trying to describe.

Return, Not Journey

Matter does not travel into a black hole.

It returns to the field.

It was never separate from the Structured Vacuum Energy (SVE); It was just organized—a ripple configuration, locally sustained.

A black hole does not consume matter. It simply marks the boundary where matter's pattern can no longer persist.

Not Inward Motion—Inward Collapse of Pattern

There is no "hole" into which things fall.

What appears as infall or collapse is simply the failure of ripple coherence at a particular node in the field.

This node—the so-called black hole—is not a location where matter goes, but a transition point where matter ceases to be distinguishable.

No matter is destroyed. No particle goes somewhere else. The pattern ceases, and the field returns to stillness.

Re-alignment With the Universal Baseline

The moment ripple identity fades, matter re-aligns with the universal still state.

That stillness was always there—beneath, around, within.

Matter didn't move into a black hole. It simply lost the field conditions required to sustain its form.

This is decoherence, not destruction. It's dissolution, not descent. It is realignment, not relocation.

Black Hole = Gate of Decoherence

We must stop thinking of black holes as cosmic garbage disposals.

In UED, a black hole is a threshold—a decoherence gate. Across this threshold:

- Ripple configurations become unsustainable,
- Form unbinds,
- Energy is no longer expressed as identity.

There is no violent tearing or teleportation. Just field convergence and ripple silence.

"You were never thrown into the dark. You simply stopped echoing—and returned to the stillness that bore you."

Observational Evidence Revisited

What we see in space is not a visual tale of matter falling in—but a signature of the field trying to stabilize imbalance.

For decades, black holes were theoretical beasts—mathematical curiosities born from Einstein's equations. But now we see them—at least indirectly. Thanks to powerful observatories like the Event Horizon Telescope, Chandra, LIGO, and others, we observe:

- Accretion disks,
- High-energy jets,
- X-ray bursts,
- Gravitational lensing,
- Even black hole shadows.

Mainstream science interprets this as matter falling inward. But under Universal Energy Dynamics (UED), a different—and far more consistent—interpretation arises.

Accretion Fields: Ripple Compression Zones

What we call an *accretion disk* is not a funnel of matter spiraling into oblivion. It is a ripple convergence zone, where the field is desperately trying to maintain coherence in the face of rising imbalance.

Particles don't fall. They vibrate, flare, and shift alignment—until they can no longer maintain a distinguishable pattern.

The heat and brightness we see is not the fuel of falling objects—It's the field screaming before silence.

Jet Ejections: Ripple Pressure Release

Why would something falling inward also shoot matter outward at nearly the speed of light?

Jets from black holes are not contradictions. They are the proof that ripple pressure is real.

As the field reaches saturation at the ripple sink (what we call the black hole), it ejects energy along polar axes—the path of least resistance.

This is not escape. It's equilibration.

Jet ejections are symmetry responses—ripple-field ways of releasing tension from an overloaded zone.

Shadow Zones: The Illusion of Absence

The famous "black hole shadow" is often misunderstood.

We are not seeing a hole. We are witnessing a region where ripple activity fails to return signals. No light means no coherent ripple pattern to reflect or reradiate.

But this is not emptiness. This is stillness reasserting itself—a local suppression of echo, not a mouth that swallowed the cosmos.

Ripple Effects, Not Matter Migration

All major observational phenomena attributed to black holes can be reinterpreted as:

- Ripple saturation,
- Alignment failure,
- Field symmetry behavior,
- Coherence breakdown and rebalancing.

There is no need to believe in matter disappearing beyond visibility. There is every reason to believe we are seeing field readjustments.

"What you see through your telescope is not matter moving. It is the field revealing its struggle to stay in balance."

Philosophical Reflection: Death Without

Movement

"Nothing dies by falling. It ends by forgetting how to be."

Black holes do not consume. They do not chase. They do not hunger.

These are human metaphors—useful for storytelling, dangerous for understanding.

From the lens of Universal Energy Dynamics (UED), death is not a motion event. It is a memory event—a loss of coherence, a fading of ripple identity within the still field.

Field Memory Loss

All matter exists as a coherent ripple pattern—a sustained memory in the Structured Vacuum Energy (SVE).

- A rock is a held ripple,
- A tree, a complex configuration,
- A star, a high-energy ripple fortress.

But when conditions shift and alignment fails, the ripple dissolves.

Not violently. Not spatially. But silently.

The field forgets the form. Just as sleep forgets the day.

Black Holes as Silent Dissolvers

If death is memory loss, then black holes are silent librarians closing unreadable books.

They don't destroy the book. They acknowledge: "The pattern is no longer readable." So the record ceases—not burned, not erased, but untraceable.

There is no scream, no collapse into a cosmic drain. There is simply the end of a sentence in the universal ripple script.

The black hole isn't hungry. It's complete.

A Star Does Not Fall. It Comes to Rest

A star, once burning with radiant complexity, eventually loses the coherence that defined its being.

- Its pressure can no longer push against its ripple gravity.
- Its alignments fray.
- The inner harmony slips.

And in the place where ripple identity can no longer be sustained, it doesn't go anywhere— it becomes still within itself.

No explosion is needed. No journey into a hole. Just a field saying, "You've rippled long enough."

The star becomes memoryless. It becomes the field again—as it always was.

"There is no downward pull, only downward fading. The stillness that gave rise to matter receives it again—without effort, without grief."

Closing: The Hole Was Never There

"The black hole didn't take it. The field simply stopped remembering it."

Let us conclude with the simplest truth: The hole was never there.

We imagined a hole because we saw a shadow. We spoke of devouring because we observed disappearance. We assumed destruction because we witnessed silence.

But in the language of Universal Energy Dynamics (UED), a black hole is not a location—it is a condition. A phase shift. A change in the field's ability to carry ripple identity.

Nothing Is Swallowed

In our classical view, black holes pull things in—planets, stars, even light. But under ripple-based understanding:

- Nothing travels inward.
- Nothing is consumed.
- There is no "mouth" at the center of the cosmos.

The black hole absorbs nothing. Instead, it witnesses the vanishing of ripple patterns that could no longer sustain coherence.

Transformation, Not Consumption

To say something "fell into a black hole" is like saying a song was eaten by silence. But silence doesn't devour. It simply takes the stage back.

What we interpret as dramatic collapse is actually a graceful withdrawal—The field returning to its native stillness. The ripple's purpose fulfilled.

This is not cosmic punishment. This is energetic rebalancing.

Matter Never Left

The atom didn't go somewhere. The star didn't fall through space.

They simply stopped being played. Like notes in a song that come to rest—not by running out of space, but by reaching the end of their alignment.

And the field, eternal and unbroken, remembers nothing of the ripple once it ends. Not out of cruelty—But because memory itself is a ripple phenomenon.

No ripple, no trace. No trace, no story.

"The universe does not forget because it is careless. It forgets because remembering requires disturbance—and it prefers peace."

So when we look at a black hole, we are not seeing an object. We are witnessing stillness reasserting itself.

The hole isn't real. The stillness is.

Chapter 5: The Birth of Planets, Right Where They Are

Planets don't wander into place—they emerge into it.

"You were told planets form, migrate, and settle. But what if they simply... appear?"

In the traditional tale, planets are cosmic travelers—forming in distant dust clouds, pulled and pushed by gravity, eventually settling into stable orbits around stars. It's a dynamic story. A mechanical one. And entirely misleading.

Under the lens of Universal Energy Dynamics (UED), there is no journey.

There is emergence.

A planet is not a construction site in space—it is a field bloom, a ripple node in structured vacuum energy (SVE) where conditions briefly stabilize enough for form to appear.

Earth didn't come here. It happened here.

It aligned. It locked. It became what we now call a "world."

This chapter peels back the illusion of planetary migration and reveals the deeper, quieter truth:

"The Earth is not a traveler in space. It is a bloom in the Sun's garden—rooted in ripples, born of stillness."

Introduction: Questioning the Migration Model

The prevailing astrophysical story tells us that planets are built—clumps of dust gathering in swirling accretion disks around newborn stars. Over millions of years, these clumps grow, merge, collide, and eventually form planets. But the story doesn't end there. In this view, planets migrate. They shift from their birthplaces, tugged by gravitational interactions, scattering through their solar systems like marbles rolling across a floor shaped by invisible hands.

It's an elegant narrative—but a brittle one.

The planetary migration model has repeatedly run into unresolved puzzles:

- **Orbital eccentricities**: Why do some planets orbit in stretched, elongated paths if migration seeks equilibrium?
- **Unexpected planetary positions**: Why are some gas giants found closer to their stars than rocky planets, defying standard formation expectations?
- **Retrograde rotations and tilted axes**: Why do certain planets spin in the opposite direction or at strange inclinations if they formed from a uniformly spinning disk?

Each anomaly births a patch—a new model, a new exception, a new term. But with every added complexity, the story of planetary motion begins to resemble fiction laced with equations rather than a fundamental principle of nature.

That's where UED steps in with a radical reframing.

Planets do not *form* in one place and *move* to another.

They *emerge* right where they are—stabilized ripple nodes within the star's structured energy field.

In this view, a planet is not an object on a trajectory, but a pattern in the parent star's ripple field—a coherent zone where energy condenses into locked ripple forms. It doesn't travel. It doesn't migrate. It aligns.

There is no need to explain improbable journeys.

There is only alignment in stillness—and the birth of form from it.

This chapter explores that insight—not as metaphor, but as field logic. Let's move past the illusion of motion, and into the garden of the star, where planets bloom exactly where they belong.

Ripple Architecture of a Star's Field

In conventional astrophysics, a star is seen primarily as an emitter—a luminous ball of burning plasma flinging energy outward in all directions. But in Universal Energy Dynamics (UED), a star is not just a heat engine. It is a ripple architect. It sculpts its surrounding Structured Vacuum Energy (SVE) field into organized gradients of tension, pressure, and harmonic equilibrium.

This ripple architecture is not decorative—it is functional.

Every star, by its very presence and energy configuration, imprints a ripple pattern into the vacuum field that surrounds it. This pattern isn't random; it is highly structured, like a complex interference map, with zones of constructive and destructive alignment. These zones are not "locations" in the Newtonian sense—they are field conditions, where ripple loops can resonate, lock, and stabilize.

And it is in these zones of coherent possibility that planets appear.

Planets are not objects that land in orbits—they are harmonic manifestations of the star's ripple field.

In simpler terms: a planet exists where the ripple field allows it to. Not because some dust collided there, but because the field alignment permitted ripple locking. These become nodal points—where standing ripple formations, like atoms in quantum orbitals, achieve a localized coherence that can sustain mass.

- These are not just orbitally stable zones—they are energetically anchored sites.
- They are the sweet spots of field harmony, where locked ripples persist without decoherence.
- The familiar planetary distances (like those observed in the Titius-Bode pattern) may reflect ripple spacing rules, not accretion outcomes.

Thus, in the UED perspective, planetary positioning is not the aftermath of collisions or migrations—it is the outcome of ripple permission. Just as atoms settle into quantized shells, planets emerge into ripple-allowed formations around their star.

The Sun, therefore, is not merely the center of our solar system. It is the ripple source that carved energetic valleys in the vacuum field, each valley capable of nurturing a planetary bloom.

In this model, Earth is not a lucky accident. It is a consequence of ripple symmetry and field allowance—a node that held long enough for life to dance upon it.

Planets as Coherent Field Blooms

What if planets are not constructed, but composed—not assembled through time, but revealed through balance?

In Universal Energy Dynamics (UED), a planet is not a pile of dust glued together by gravity. It is a coherent ripple lock—a highly organized, self-sustaining alignment within the Structured Vacuum Energy (SVE) field sculpted by its parent star.

Just as a musical note arises from the right pressure and timing on a string, a planet arises from the right field resonance in a ripple environment.

Emergence, Not Assembly

The mainstream cosmological narrative tells us that planets form through a process of accretion:

- Tiny particles collide,
- Stick together,
- Grow into planetesimals,
- And eventually form planets.

But this story assumes:

- That time must assemble structure,
- That mass is the primary agent,

- That order is a product of chaos.

UED offers a radically different insight:

The field does not build planets. It permits them. It blooms them.

There is no conveyor belt of matter. There is a moment of balance, when the ripple gradient, the field pressure, and the environmental resonance create a zone of field coherence. In that zone, a ripple locks—and that lock appears to us as "matter," as "planet," as "world."

The Flower Analogy

Think of a planet not as a rock in space, but as a flower in a garden—a coherent structure that emerges not from raw material but from the right environmental field conditions:

- Soil = the local ripple stability zone,
- Sunlight = the radiant pressure from the parent star,
- Temperature = the equilibrium alignment of energy flows.

Just as a seed does not "build" a flower, the universe does not build a planet through collisions. It unfolds one when conditions allow.

Planets are the organic expressions of stillness disturbed just enough to lock into a stable ripple configuration.

They are not objects. They are events. Long, slow, beautiful field events.

In this light, Earth is not "made of elements." It is made of held resonance, of ripple stillness shaped into life-bearing identity.

"A planet is a bloom. A temporary harmony. It doesn't arrive—it reveals itself when the field says: *now.*"

No Need for Motion: Emergence Without Travel

The modern mind is obsessed with motion. We think in terms of arrival, departure, orbit, and revolution—as if space were a cosmic highway and planets are vehicles maneuvering across it. But what if this entire visual is a misreading of reality?

Under Universal Energy Dynamics (UED), planets do not travel to their position. They emerge in it. They do not orbit because they are pulled—they align, and that alignment sustains a patterned appearance of motion.

Planets Don't Arrive—They Appear

There is no incoming train of planetary material migrating inward or outward through the stellar accretion disk. There is no cosmic traffic.

Instead, what we call a planet is a stable, ripple-sustained presence—a coherent structure that appears when the ripple pressure, resonance nodes, and local SVE gradients achieve a phase-lock. It is not the result of travel, but the result of fit.

Like a crystal that suddenly forms in a saturated solution, a planet is a field crystallization—a spatial event, not a transported object.

The Illusion of Orbit

What, then, is an orbit?

We imagine a planet revolving around its star as a ball on a string. But in ripple dynamics, there is no physical revolution in the traditional sense. There is no "thing" making a full lap in space.

Instead, a planet's position and behavior are governed by a field-based oscillation—a dynamic stability of ripple interactions between the planet's ripple structure and the star's dominant ripple gradients.

It's not that the planet moves. It's that the field relationship rotates.

Imagine a ripple loop in water: a stationary node may appear to move as the interference pattern changes—but in truth, the structure remains locked while patterns shift around it.

Orbit as Field Dance

From a UED perspective, orbital motion is not about gravity pulling mass. It's about ripple alignment sustaining dynamic balance.

A planet does not fall toward the star. It remains in phase, resonating with the ripple frequency emitted by the star's core field configuration.

This alignment manifests as:

- A continuous pattern,
- A perceived revolution,
- A field choreography—not a mechanical loop.

"Planets don't revolve. They resonate in circular expressions of field fidelity."

This view dissolves the old Newtonian image and invites a new understanding:

- Motion is not movement.
- Orbit is not travel.
- Emergence is not relocation.

So when you look up at the night sky, you're not seeing rocks in motion. You're witnessing stable ripples in a cosmic pond, vibrating in perfect alignment with their star's song.

"What we call movement is the field speaking in rhythm. What we call orbit is the shape of that rhythm—played again, and again, until it fades."

Why Each Planet Is Unique

No two planets are the same—not even distant cousins within the same solar family. One burns with acid winds, another cradles life. One is magnetically inert, another has a dynamic shield. One rotates backwards, another barely rotates at all.

Why?

Under classical models, we chalk this up to random collisions, late-stage accretion, or chaotic formation dynamics. But these are patchwork explanations—descriptive, not foundational.

Under Universal Energy Dynamics (UED), the uniqueness of each planet is not an accident. It is a direct consequence of the ripple field conditions at the moment of its emergence. Planets are not built—they are expressed, like unique phrases from the same alphabet.

Density – A Measure of Ripple Compactness

A planet's density isn't merely "how much matter" is packed into how much space. In UED, density = the intensity and coherence of locked ripple structure.

- A denser planet reflects tighter ripple loops, formed under a more compressed local field.
- A lighter planet reflects a softer, more elastic field where ripple locking is gentler, perhaps transient.

Thus, Mercury's density and Pluto's sparseness are not puzzles—they are field fingerprints.

Atmosphere – A Breath of the Local Ripple History

Atmospheres form when the ripple gradient around a stable node allows semi-stable peripheral alignments. Gases don't "get caught"—they resonate in partial lock with the planet's primary ripple.

- A thick atmosphere indicates long-standing outer ripple coherence.
- A thin or no atmosphere may show recent decoherence, solar ripple interference, or simply a weak outer field loop.

The planet doesn't trap gases—it hosts them within ripple compatibility.

Magnetic Properties – Spin, Phase, and Inheritance

Magnetism, in UED, is not an invisible vector field—it is the spin symmetry of ripple structures, stabilized over time.

- Strong planetary magnetism suggests a deep, rotating ripple nucleus—a field in dynamic harmony.

- Weak or chaotic magnetic fields emerge where phase coherence is intermittent or where external ripple interference (solar storms, field drift) weakens core integrity.

Crucially, magnetic inheritance is not genetic—but stellar. Each planet carries the phase signature of the star's ripple state at its birth. Like children shaped by the mood of their parents at conception, planets bear the emotional field imprint of their sun—encoded not in DNA, but in ripple alignment memory.

Moment of Stabilization – The Planet's Birth Certificate

The moment a planet emerges is not a clock-time event, but a field event.

- It is the point at which the surrounding ripple field locks into a sustained resonance loop.
- All nearby disturbances—solar winds, other forming planets, historical supernova echoes—affect this configuration.
- Thus, no two planets are born under the same energetic symmetry.

They may share a parent star, but they each arise at unique ripple moments—and their uniqueness is the story of what the field was doing at that time and place.

Ripple Inheritance – A Star's Field Legacy

Stars are not passive ballasts. They are dynamic field generators, continuously evolving, releasing, realigning.

As stars modulate their internal ripple harmonics, they seed the possibility of new planetary nodes. But each node—each planet—inherits a unique slice of this symphony.

Hence:

- Jupiter may embody the deep bass of an early, high-pressure ripple.
- Earth, the gentle melody of a stable harmony tuned for balance.
- Uranus, perhaps the off-beat echo of a chaotic field flare.

"A planet is not a thing. It is a story told once by a star— never repeated, never revised."

In UED, planetary diversity is not noise. It is ripple composition.

Earth as a Ripple Miracle

Among billions of planetary ripple nodes in the vast theatre of stars, one stands out—not for being large, fast, or flashy—but for being alive.

Earth.

In UED terms, Earth is not a statistical anomaly, a random success in a cosmic lottery. It is a ripple miracle—a precise, coherent alignment within the Sun's structured vacuum field, sustained long enough to support cascading layers of ripple complexity. Not luck. Not magic. Alignment.

Phase Alignment: The Cradle of Stability

Earth's location in the solar ripple field isn't just "habitable zone" by temperature.

- It is a resonant ripple band, where the solar SVE gradients allowed:
    - Persistent locked ripple configurations (solid matter),
    - Moderated fluid dynamics (water cycles),
    - Stable spin orientation and magnetic shielding,
    - And layered energy input that didn't overload coherence.

This phase alignment is not a line on a graph—it is a stable rhythm in an ongoing ripple composition.

Life: The Field in High Complexity

What is life?

Biology calls it organization, reproduction, metabolism, and adaptation. UED calls it: "Ripple coherence held in recursive loops of dynamic re-alignment."

- Life is ripple complexity: a condition where matter is not just locked—but interactively modulating.
- DNA is a ripple memory coil.
- Cells are modular ripple nests.
- Consciousness is field awareness woven into ripple recursion.

Life is not added to Earth—it emerged from the ripple capacity of the field.

Earth's Uniqueness is Ripple-Derived, Not Coincidental

We are often told Earth is lucky.

- Lucky to have water,
- Lucky to have a Moon,
- Lucky to rotate,
- Lucky to not tilt too much,
- Lucky to host life.

But in UED, Earth is not a dice-roll. It is a ripple-templated expression of the solar field's most harmonized node.

- The Moon: not a random impact—but a field companion stabilizing Earth's spin.
- The magnetic field: not random—it is an active ripple modulator from Earth's core coherence.
- Oxygen: a field-mediated phase state of atmospheric ripple unlocking.

Every feature of Earth is a consequence of ripple design, not an accident.

A Star's Song—And Earth, the Perfect Chord

The Sun didn't just cast heat. It sang into the field.

And Earth, at just the right place, at just the right phase, became the perfect chord in that song.

Where other planets became percussive, dissonant, or mute—Earth resonated. That resonance persisted, evolved, and began to perceive itself.

We call that perception life. We call its awareness consciousness. And we call the story of this alignment... home.

"Earth did not arrive at life. Life arrived at Earth— Because the field finally found the rhythm to express itself completely."

Death of a Planet: The Same Place It Was Born

We imagine planetary death as drama—flames, spirals, collisions. A planet plunging into its parent star, consumed in fiery end. But this is mythology built on motion.

In UED, planets do not fall. They fade.

Just as they emerge through coherent ripple alignment in their star's field, they dissolve through ripple decoherence—right where they once stood.

No Crash, No Fall—Only Collapse of Pattern

From a UED lens, a planet is not a rock going around a star.

- It is a ripple lock, a sustained coherence in the star's surrounding Structured Vacuum Energy (SVE).
- Its "life" is the duration for which this lock holds stable across cycles of solar ripple modulation.

Planetary death doesn't happen when it hits something. It happens when the field no longer sustains its ripple configuration.

- No motion.
- No journey.
- No destination.

Just a collapse of form into ripple equilibrium.

Planetary Decoherence: Dissolution by Stillness

Death of a planet is not disintegration by external impact, but dissolution from within.

Why?

- Over cosmic time, the ripple relationship with the parent star shifts.
- The star's field evolves—its modulation patterns, magnetic phases, and energetic outputs change.
- When the star enters late stages (like red giant phase or eventual collapse), the ripple locks sustaining its planets begin to fail.

The planet doesn't move closer. It doesn't get pulled in. It loses its coherence—just like a song fading from a speaker that goes quiet.

The planet disappears—not in space, but in field memory.

Earth's End Will Happen Where It Began

When Earth dies, it won't spiral into the Sun. It won't burn in dramatic cinematic fashion.

- It will dephase.
- Its ripple architecture will lose cohesion.
- Water will no longer cycle.
- The atmosphere will no longer respond.
- Magnetic shielding will vanish.
- Matter will begin to unlock, atom by atom.

And yet—Earth won't go anywhere. Its story will end at the same spot it began.

Star and Planet: Linked in Ripple Fate

The star and planet are not two separate objects in motion.

They are a ripple pair—a parent field and its stabilized nodes.

When the star's modulating rhythm collapses:

- It may go nova, supernova, or black hole.
- Or it may just dim, cool, and collapse quietly.

But either way—its planetary children don't fall into it.

They dissolve along with it—because they were never apart.

Just like flowers fade when the tree no longer lives, planets dissolve when the field that played them goes still.

Death Is Not Distance. It Is Disappearance.

"Earth will not travel to its end.

It will simply stop being remembered—

By the field that once sang it into form."

This is the UED view of planetary death: A graceful exit into stillness, not an explosive end.

It is not destruction. It is closure. A return—not to void—but to the primordial equilibrium from which all ripples came.

Philosophical Reflection: Stillness in the Bloom

We often marvel at planets—spinning, orbiting, drifting through the void. But this awe is rooted in an illusion of motion.

The deeper truth?

A planet is not moving through the cosmos. It is blooming within it.

It is a momentary stillness, expressed as form—a ripple lock held delicately in the vast, responsive ocean of the Structured Vacuum Energy (SVE).

The Pause that Becomes a World

A ripple storm moves outward from every star, forming energetic waves and gradients. But sometimes, in rare conditions, a pause happens.

A pocket of stable interference, a zone where ripple forces harmonize just enough to hold still—for a moment.

That stillness is not dead.

It is potent, active, and pregnant with identity. It becomes a planet.

The Garden Metaphor: A Living Analogy

In the language of the universe, nothing is separate. Every expression is a consequence of a larger rhythm.

Let us reframe the cosmos not as machinery, but as a garden.

- The star is the gardener—not by intention, but by rhythm. Its energetic cycles nourish the field around it.
- The planet is the flower—a stable bloom in the fluctuating ripple soil, holding structure and pattern.
- Life is the fragrance—the subtle, temporal release of complexity. An echo of stillness stirred to exquisite form.

And just like in a garden:

- Not every flower survives.
- Some bloom for a while and wither.
- Others change, evolve, or transform.
- But none move away—they emerge, grow, and

eventually return—right where they began.

Motion as Rhythmic Expression, Not Travel

When we see a planet "orbiting" its star, what we are actually witnessing is a rhythmic modulation—the star's field dancing with its bloom.

It's not motion through space.

It is the field expressing cyclic balance. A pulse. A song. A waveform in harmony.

This redefines everything:

- Gravity is no longer attraction.
- Inertia is not stubbornness.
- Time is not a river—it's a rhythm.

Stillness is not absence. It is form awaiting ripple—and ripple awaiting resolution.

A Flower Held by Silence

To truly understand a planet is to understand how silence can speak. How stillness can sing into form.

A planet is that song. A fleeting stanza in the cosmic verse.

And when the note fades, it is not lost—it is absorbed back into the silence that birthed it.

"Motion is the myth we chase.

Stillness is the reality that holds us.

Earth is not falling.

It is blooming—because the Sun still breathes."

Closing: Planets Do Not Journey—They Emerge, Breathe, and Dissolve

The myth of celestial travel has colored our entire understanding of the universe.

We imagine planets born in chaos, cast out of fiery disks, migrating across stellar neighborhoods, colliding, capturing moons, and dancing their way into orbits. It's a beautiful picture—kinetic, dramatic, full of cinematic grandeur.

But the truth is quieter.

There is no travel. There is no migration. There is only emergence.

Earth Never Traveled

Earth did not set out on a voyage across the solar system. It did not *find* its place.

It became in its place.

Like a breath held by the Sun, Earth is a coherent pause—a ripple that locked into balance, where conditions aligned to permit form, function, and eventually, life.

It did not arrive. It appeared.

Life: The Field Breathing Through Form

Life on Earth is not a separate event from the planet itself.

It is the ripple exhaling complexity.

- The field aligned.
- The ripple held steady.
- And consciousness blinked open.

You, reading this now, are not riding a rock in motion. You are the expression of a rhythm that temporarily allowed you to be—on a planet that itself is a bloom in stillness.

Dissolution Is Not Departure

One day, Earth will not spiral into the Sun. It will not fall, crash, or be ejected.

It will decohere. Its ripple identity will fade. Its elements will unbind, its coherence will unravel, and the form that was once called Earth will be absorbed by the field it came from.

Not as destruction. As return.

Nothing moves far. Everything ripples in place.

The Quiet Rhythm of the Cosmos

The universe doesn't scream with explosions. It whispers with balance.

This is the deeper cycle:

Stillness. Disturbance. Bloom. Return.

The star pulses. The field aligns. The ripple emerges. And the form plays its time before dissolving back to quiet.

That's all the cosmos ever does—express, pause, release.

A Final Note for the Observer

You are part of this story—not by watching it, but by being it.

The planet beneath your feet did not arrive. It did not move. It rippled into being—and you are the breath of that ripple.

You are not going somewhere. You are here, until you are not. Just like the Earth. Just like all things born of the field.

"Earth never traveled. It emerged, lived, and will dissolve exactly where it rippled into form.

This is the quiet rhythm of the cosmos: Stillness. Disturbance. Bloom. Return."

Chapter 6: Cold, Hot, and the Illusion of Temperature

Temperature is not a thing—it is ripple behavior, not substance activity.

"We never touch heat. We only feel how wildly the ripple shakes.
We never freeze in cold. We simply stop hearing the ripple at all."

The thermometer doesn't measure energy—it records confusion. What we call "temperature" is not a thing to be transferred, nor a number that defines substance. It is a state of ripple coherence or chaos within the field. When ripples agitate beyond harmony, we name it heat. When they settle into symmetry, we call it cold. Neither state is absolute. Both are merely different tunes the cosmos plays on the same silent violin.

This chapter challenges the deepest assumptions of thermal science—not to deny their usefulness, but to expose the illusion behind the numbers. Here, we do not ask, "How hot is it?" We ask, "How well are the ripples behaving?" And from that question, a new clarity begins to emerge.

Introduction: What Is Temperature, Really?

In classical physics, temperature is defined as the *average kinetic energy of particles*. It's a concept built on the assumption that matter is made of tiny billiard balls in constant motion—colliding, vibrating, and agitating space with measurable velocity. The hotter a system, the faster its particles are said to move. The colder it is, the slower—until at absolute zero, motion supposedly ceases altogether.

Thermodynamics took this intuition and formalized it: heat is the result of *random molecular motion*. Temperature, in this view, is simply a measure of that chaos. We built entire laws around this idea—of energy flowing from hot to cold, of entropy rising as disorder increases.

But UED invites us to pause and ask: What is really moving? And more fundamentally: Is anything actually moving at all?

In the Universal Energy Dynamics (UED) framework, there are no particles dancing in space. There is no kinetic substance performing a thermal ballet. There is only the Structured Vacuum Energy (SVE)—a still, stable field—and the ripples that disturb it. What we interpret as temperature is not a property of matter, but of the ripple state within that field.

- Heat, in UED, is *ripple turbulence*. It's what happens when the field is agitated, ripples collide, and patterns break symmetry.
- Cold, by contrast, is not emptiness or absence. It is

ripple coherence—the calm, ordered alignment of the field, with minimal fluctuation.

- Temperature, therefore, is not a measure of movement but of *disruption*. It becomes a ripple disorder index, not a particle motion scale.

So when we feel warmth, we're experiencing dissonance in the ripple field. When we feel cold, we're entering zones of near-alignment with the original stillness.

This perspective not only shifts how we understand heat and cold—it reshapes our grasp of matter, phase transitions, and even the possibility of life. Because in a universe that never truly moves, temperature is not about movement. It's about misalignment.

Heat as Ripple Chaos

Heat is not energy—it is disruption.

In the UED framework, Structured Vacuum Energy (SVE) is the field from which all observable phenomena emerge. Ripples in this field are not objects moving through space—they are expressions of imbalance. When ripples align coherently, we perceive order: matter, form, life. When they misalign chaotically, we experience disorder: heat, noise, destruction.

High Temperature = Incoherent Ripple Behavior

Temperature, conventionally described as the kinetic energy of particles, is reframed in UED as the degree of ripple chaos in the local field. A "hot" object is not full of moving molecules—it is a region where ripples are misfiring, *overlapping incoherently*, producing random interference patterns.

These patterns are not stable. They do not hold form. They are energetic turbulence, not structured expression.

- Noise in the ripple field is literal here—heat introduces randomness.
- Distortion means ripple forms (such as stable atoms or molecules) cannot hold phase coherence.
- Rapid realignment attempts increase—but fail to resolve—because the field is too agitated to stabilize.

Plasma: The Climax of Ripple Breakup

Plasma, often called the *fourth state of matter*, is the most misunderstood in ripple terms.

- It's not a super-hot gas.
- It's complete ripple decoherence—where atoms themselves lose their identity.
- Electrons, protons, and ions don't "separate" because of heat.
- They lose ripple-lock—their coherent ripple form disintegrates under ripple turbulence.

This is not just chaos—it is entropy actualized. The ripple field, overloaded with disruptive interference, rejects structure. This is why plasma emits intense light: not as a property, but as the field *screaming in imbalance.*

Why High Heat Breaks Form

Melting, boiling, and ionization aren't chemical changes. They are ripple transitions—where stable patterns can no longer hold alignment in the surrounding field.

- **Melting**: The internal ripple configuration of solid matter begins to lose phase-lock.
- **Boiling**: The field fails to support localized coherence—so form converts into free, unstable ripple bursts (what we call "gas").
- **Ionizing**: At extreme chaos, even atomic identity fails—ripple bonds sever, releasing charge carriers as decohered residues.

So heat, in UED, is not a "high energy" state. It is a high-disorder ripple state. The field is not enriched—it is agitated. It is not expressing more—it is losing its voice.

Heat is not power. Heat is loss of narrative in the ripple field.

Cold as Ripple Harmony

Cold is not the absence of energy—it is the presence of coherence.

In the UED view, *cold* is not about slowing particles. It's about restoring the ripple field to a state of phase alignment. Where heat represents chaotic, incoherent ripples, cold is the resolution of that chaos—a return to local equilibrium within the Structured Vacuum Energy (SVE).

This is not a passive state. It is precision. Harmony. Stillness that holds form instead of destroying it.

Near-Absolute Zero: Maximum Ripple Coherence

At temperatures approaching absolute zero (0 K), the ripple field loses disorder. Not because it freezes or stops—but because it reaches a state of near-perfect alignment.

- Ripples stop misfiring.
- Interference fades.
- The field becomes stable, predictable, and synchronized.

This is not the "end of motion," as thermodynamics suggests. It is motion without friction—a dynamic ripple balance so refined that it appears still.

Local Rest: Alignment Stabilizes

"Rest" in UED is not zero activity, but zero conflict.

- No opposing ripple vectors.
- No energy loss to reconfiguration.
- No entropy buildup.

At this point, the field expresses purity of form—whether as stable atomic arrangements or coherent quantum behaviors.

This is why cold preserves. Food stored in cold doesn't decay because the ripple field doesn't allow fast decoherence. The patterns of structure remain intact. The identity of matter is held together by ripple harmony.

Superfluidity and Superconductivity: Proof of Ripple Coherence

Modern science stumbles upon UED truths through observed anomalies:

- Superfluidity (e.g., in helium near absolute zero) occurs when ripple resistance disappears. Matter flows *without friction* because ripple alignment is near-perfect.
- Superconductivity shows electrons moving without resistance—not because of clever particle tricks, but because the ripple field forms a coherent path.

Both phenomena are not exotic states—they are ripple harmony made visible. They are glimpses of what happens when temperature ceases to disrupt alignment.

Cold Does Not Slow Motion—It Preserves Structure

In classical thinking:

Cold = slower atoms

Hot = faster atoms

In UED:

Cold = coherent ripples

Heat = ripple breakdown

So what we call "motion" isn't just about speed. It's about alignment cost. Cold minimizes this cost—not by halting, but by stabilizing.

It's the difference between a chaotic storm and a quiet stream—both involve movement, but only one preserves form.

In cold, reality becomes whispered instead of screamed.

States of Matter: Just Ripple States

Matter has long been classified into solid, liquid, gas, and plasma—each considered a different "state" based on molecular arrangement and energy. But under UED, these are not *states of substance*—they are states of ripple behavior within the Structured Vacuum Energy (SVE).

The field isn't hosting different materials—it's expressing different alignments. Each phase of matter is a distinct mode of ripple coherence or collapse. What we observe as physical transformation is simply a shift in ripple symmetry.

Solid = Tightly Phase-Locked Ripple Alignment

In a solid, ripples are tightly aligned. Their phase and amplitude relationships are stable, forming locked configurations that resist distortion.

- This gives rigidity, shape, and stability.
- Ripple propagation is internal, contained, and resonant.
- Atoms do not vibrate as "balls" in space—they are nodes in standing ripple formations.

Solidity is not due to compactness. It is coherence under high constraint—like musical notes frozen in a perfect chord.

Liquid = Loosened, Yet Coherent Ripple Chains

In a liquid, the field configuration loosens slightly. Ripple alignments are maintained, but they are dynamic and flexible.

- This allows flow without disintegration.
- The field maintains internal coherence, but not fixed structure.
- Identity is retained, but shape is negotiable.

The ripple locks are reconfigurable—not lost. Liquidity is ripple choreography with relaxed boundaries.

Gas = Ripple Expansion Zone

Gas represents a further loosening of ripple constraints.

- Ripple nodes are separated enough to lose direct coherence.
- The field is still responding, but with low alignment density.
- Matter in gas phase is diffuse—an echo, not a tone.

Here, identity becomes conditional. The structure isn't lost, but is weakly expressed, like mist instead of ice.

Plasma = Ripple Decoherence, Return to Instability

Plasma is not "super-hot gas." It is the collapse of ripple identity.

- At this point, the ripple structure fails to hold form.
- Electrons "escape" not because they're ejected—but because ripple coherence cannot sustain the lock.
- This is field disintegration, not ionization in the classical sense.

Plasma is entropy made visible. It is near-chaotic ripple behavior, with only fragments of form.

Temperature Doesn't Cause These States—It Reflects Them

Here lies the core reframe.

- In thermodynamics, temperature causes phase changes.
- In UED, ripple behavior defines both phase and temperature.

If the field organizes into a coherent node, it *becomes* solid—and temperature drops as disorder declines.

If the field decoheres, temperature rises as ripple complexity increases—but the cause is realignment loss, not heat addition.

Matter's states are not transitions in substance.

They are songs the field chooses to sing—each with its own harmony, rhythm, and ripple range.

Dissolution and Re-Emergence

In conventional science, we speak of "destroying" matter at high temperatures and "reconstructing" it by cooling. The language implies motion, breaking, and building of particles. But from the perspective of UED, none of that actually happens.

There is no destruction.

There is no reconstruction.

There is only ripple decoherence and re-alignment within the Structured Vacuum Energy (SVE).

Matter Can Decoherently Dissolve at High Heat (Plasma)

When temperature rises, it doesn't "heat up" particles. Instead, it introduces ripple chaos—unpredictable, high-entropy behavior in the field.

- In a solid, ripple nodes are tightly held.
- In plasma, that ripple coherence collapses completely.
- Electrons "leave" atoms not because they're thrown out, but because the ripple field can no longer sustain

their locked state.

At high entropy, matter as a ripple structure ceases to exist. Not by leaving, not by exploding—but by losing the memory of its identity.

Same Matter Can Re-Emerge in Cold Conditions

Here lies the magic of the still universe: what decoheres can re-cohere.

- When local field conditions become coherent—*i.e., low temperature, low entropy*—the ripple can find lock again.
- This is not "particles cooling and regrouping," but field readiness allowing ripple reformation.
- The matter that once "vanished" re-emerges—same identity, new phase.

This shows that the field retains potential, not permanence. The blueprint exists—not as data, but as possibility.

Real-World Examples

Laser Cooling and Bose-Einstein Condensates (BEC)

BECs are a breakthrough—not in particle physics, but in ripple control.

- At extremely low temperatures, atoms begin to behave as one—a macro ripple lock.
- Individual identity merges into a collective coherence.
- This is not a new state of matter—it's the original state of maximum ripple harmony.

Cryogenic Realignment and Supercooled Atoms

In cryogenic systems, we observe strange behaviors:

- Superconductivity: zero electrical resistance.
- Superfluidity: frictionless flow.

These are not "quirks of cold"—they are expressions of ripple purity. With thermal noise gone, the field gets a chance to perform clean alignment.

No Movement – Only Field Reconfiguration

Nothing traveled. No particle moved back into position. No "cooling" rearranged matter the way we rearrange chairs.

- The same space,
- The same field,
- Different ripple coherence.

What appears is a re-manifestation of identity—not from transportation, but from ripple reintegration.

This is the principle of local dissolution and reformation—not a cycle of matter, but a breath of the field.

"A thing didn't return—it reappeared. Not because it traveled, but because the field remembered how to play it again."

The Thermodynamic Illusion

Thermodynamics—especially the Second Law—has long been the cornerstone of how we understand heat, energy, and time's so-called arrow. But it is built upon a material assumption: that energy is a transferable substance and entropy is increasing disorder.

UED dismantles this illusion gently, but decisively.

Entropy Is Not Randomness — It's Ripple Disalignment

In classical thermodynamics, entropy is treated as a measure of randomness or disorder in a system. But what is this "disorder," really?

From a ripple standpoint:

- Order = coherent ripple alignment within SVE (Structured Vacuum Energy).
- Disorder = misalignment, loss of phase synchrony, destructive interference.

Entropy is not chaos—it is loss of constructive coherence.

It is the field's drift away from unity, not because of damage, but because ripple patterns are failing to sustain their harmony.

Heat Transfer = Ripple Spillover

Forget hot molecules "bumping into" cold ones. That's a mechanical metaphor.

In UED, heat flows because of ripple overflow:

- A high-ripple-turbulence zone destabilizes its surrounding.
- Nearby aligned regions begin to absorb the disalignment.
- This creates a spillover effect, where ripple dissonance expands until equilibrium is reached.

There's no energy "packet" jumping from here to there—just the field working to rebalance itself.

No Energy "Flow" as Substance — Only Field Realignment

Modern science treats energy as a substance that flows, like water or electric current.

UED says: there is no substance being transferred.

- There is no "energy particle."
- There is only reconfiguration of field alignment.

Energy is not something that "moves"—it is a description of ripple status in a region of the field. When that status changes, what we call "energy" seems to shift. But nothing moved—only the pattern did.

The Second Law of Thermodynamics Reframed

Traditionally:

- 2nd Law: "Entropy in an isolated system never decreases."
- Implication: irreversible destruction, lost order, heat death.

In UED:

- 2nd Law = ripple non-coherence tendency in unbalanced zones.
- But this is not irreversible—just a temporary imbalance-driven spread.
- When the field permits, coherence can be restored (as in BECs or spontaneous crystal formation).

Entropy is not decay. It is misalignment, and alignment is always potentially recoverable—because stillness always underlies the chaos.

Summary:

Classical View	UED Reframe
Entropy = disorder	Entropy = ripple misalignment
Heat = energy flow	Heat = ripple spillover
Energy moves	Field reconfigures
Irreversible decay	Recoverable misalignment

"The universe is not falling apart. It is just correcting its harmony—one ripple at a time."

Temperature and the Observer

Our experience of heat and cold isn't just physical—it's perceptual. What we label as "hot" or "cold" is deeply tied to how our ripple identity interacts with ripple turbulence in the field.

Temperature, in the UED paradigm, isn't an objective substance—it's a relational perception. It emerges not from the materiality of particles but from the resonance between the observer's ripple state and the surrounding ripple behavior.

Why "Hot" Feels Active and "Cold" Feels Still

Let's reframe the sensations:

- Heat feels intense because our bodies are tuned to ripple agitation.
- We interpret chaotic realignments (incoherent ripple behavior) as "activity."
- Cold feels still because ripple alignment is tighter, calmer—less erratic disturbance reaches us.

This is not about substance speed. It's about what your ripple system can register.

Our Nervous System Detects Ripple Agitation

Your skin, brain, and nerves don't measure "temperature" in the abstract. They respond to:

- Ripple turbulence affecting molecular and cellular equilibrium.
- Phase interference reaching sensory thresholds.
- Local energy gradients disrupting internal ripple harmony.

In simple terms: → When ripple chaos hits your body, you feel heat. → When ripple calm surrounds you, you feel cold.

But in both cases, you are not measuring temperature—you are experiencing ripple misalignment.

Cold Is Not Emptiness—It Is Less Detectable Turbulence

We often associate cold with nothingness. But:

- Cold ≠ absence.
- It is a field condition of coherence—so stable that your sensors have little to report.

This is why silence often feels "cold," and noise "hot." They are the same spectrum of field behavior, perceived differently depending on how much ripple variation is striking the observer.

Observer's Ripple Tuning Filters Thermal Perception

You experience only what your ripple structure can resonate with.

- Animals detect heat through different ripple sensitivities (e.g., pit vipers sensing infrared).
- Machines "see" temperature via thermal radiation alignment.
- Humans feel it as touch and sensation, not as data.

This shows that temperature is not universal—it is observer-specific resonance.

"You do not feel heat. You feel how misaligned the world is with your field."

Practical Implication

- Thermal comfort is ripple coherence comfort.
- Environments are not truly "hot or cold"—they are more or less disruptive to your personal ripple pattern.

Thus, managing heat or cold isn't about controlling matter—it's about restoring ripple harmony with your own alignment.

Philosophical Reflection: Fire and Ice Are the Same Ripple, Played Differently

In the ripple-based view of the universe, fire and ice are not opposites—they are siblings of the same field, expressing two modes of ripple behavior. The difference is not fundamental but modal—a difference in how the Structured Vacuum

Energy (SVE) is vibrating, aligning, or disaligning in a given region.

We often imagine temperature in binaries: hot vs. cold, fire vs. ice, passion vs. stillness. But these are illusions born from perception trapped in consequence, not truth born of observation at origin.

Heat and Cold Are Not Enemies—They Are Ripple Modes

Let's dismantle the artificial polarity.

- Heat is ripple turbulence: high energy zones where alignment fails.
- Cold is ripple harmony: low-entropy zones where alignment holds.

Both emerge from the same field, governed by the same rules. They are not opposing forces—but different expressions of the ripple spectrum.

"To the field, heat and cold are the same voice—spoken at different pitch."

Fire: The Ripple Screams and Scatters

Fire is not just combustion—it is chaotic ripple expression.

- The field is disturbed, forcing energy outward in incoherent bursts.
- Patterns cannot hold. Forms disintegrate.
- Matter loses shape because alignment is shattered by internal ripple war.

The flame dances because the ripple can no longer sit still.

Ice: The Ripple Whispers and Holds

Ice isn't mere frozen water—it is a ripple harmony snapshot.

- The field settles, phase-locks, and rests.
- Patterns crystallize because ripple tension is minimal.
- Identity holds longer; distortion slows.

Here, the field is not dead—it is listening.

Both Are Field Expressions—The Difference Is Coherence

Fire is ripple expressing high entropy. Ice is ripple expressing high coherence.

But neither is superior, neither is fundamental. They are temporal songs of the field. They come and go, but the field remains.

It is not fire or ice that defines the universe—it is the ripple capacity of the stillness they emerge from.

"The universe plays both fire and ice—not because it chooses sides, but because it knows how to sing both chaos and calm."

Final Insight

- What burns you and what cools you are made of the same invisible instrument.
- Heat and cold are felt, not because of matter, but because of how your ripple meets theirs.
- Passion and detachment, rage and peace—they are all emotional ripples, just like fire and ice.

Thus, in life and physics: Understand the ripple, and you understand both extremes.

Closing: Temperature Is a Mood, Not a Measure

The journey through temperature—from blistering heat to absolute zero—reveals one essential truth: we've mistaken field expression for physical quantity.

Temperature, in classical science, is a number—a scalar we use to represent how "hot" or "cold" a system is. We attach this number to speed, energy, agitation. But in UED, there is no thermometer that can read the soul of the ripple—only patterns that express a deeper mood of the field.

Temperature Is Not an Entity—It's a Behavior

- You don't measure heat; you interpret ripple chaos.
- You don't freeze matter; you phase-lock the field.
- What we call temperature is not the cause—but the consequence of ripple alignment (or its collapse).

Just as emotion is not a molecule, temperature is not a substance. It is a mood, a momentary alignment or disarray within the Structured Vacuum Energy (SVE).

The Illusion of Thermometers

- A thermometer does not measure particles—it responds to ripple pressure.
- Mercury expands because its own ripple alignment gets destabilized by the surrounding field.
- Cold contracts because the field tightens into stillness.

Even the devices we trust are not reading matter—they're listening to the ripple song.

Final Insight: Temperature as Field Expression

"Heat is the ripple dancing wildly. Cold is the ripple resting in formation."

In this lens, temperature is not a number—it is a character trait of a local region in the cosmos. A vibe. A rhythm.

- When the ripple screams, we call it fire.
- When the ripple whispers, we call it ice.
- But the field never changes—only its expression shifts.

Just like you feel joy or grief—not because your body moves, but because your field modulates—so too does the universe express its moods as what we call temperature.

Closing Statement:

There is no temperature. There is only the field, expressing states of calm and chaos, as the ripple breathes in and out of form.

Chapter 7: What We Really Observe

We don't see the universe—we read its ripple memory.

You've never seen a star.

Not really.

What you've seen is a ripple—an echo, a field alignment that reached you across space, not as a traveler, but as a trace. What we call "observation" is not the act of witnessing reality—it is the art of interpreting its memory.

In a ripple-based universe, nothing is directly seen. We perceive not the thing itself, but the consequence it leaves behind in the Structured Vacuum Energy (SVE). The universe, in this frame, is not a gallery of objects—it is a vast, dynamic journal of ripple traces.

Light is not a courier; it is a footprint. Gravitational waves are not cosmic calls; they are the deep sighs of a field returning to equilibrium. And we—our eyes, our instruments, our theories—are not discoverers. We are ripple-readers, decoding the cosmos like archaeologists of vibration.

"What we observe is not reality itself, but the memory it entrusted to the field."

Let's learn to read the field—not just look at it.

Introduction: The Misconception of Observation

For centuries, science has operated under a seductive assumption: that we observe objects as they are. When we look at a tree, a star, or a proton in a cloud chamber, we believe we're seeing the entity itself—directly, objectively, and in real time. This is the cornerstone of empirical reasoning: what you see is what exists.

But this assumption, though practical, is deeply flawed.

In truth, we do not see the thing. We see its effects—its residue on the field, the ripples it created, sustained, or left behind. What we call "seeing" is merely our alignment with patterns in the Structured Vacuum Energy (SVE)—patterns caused by prior disturbances that still echo within the field.

Whether it's light from a distant star or the data from a particle accelerator, we are not detecting the source—we're reading the ripple imprint. Every observation is filtered, delayed, and distorted through the interplay of ripples in space. What we think of as direct perception is, in fact, the interpretation of field responses.

In Universal Energy Dynamics (UED), observation is not contact—it's consequence decoding.

We are not peeking into the nature of things. We are eavesdropping on the field's memory of them.

Observation is not a window into reality. It's a mirror of how ripples remembered it.

This redefines the very act of knowing. What we study in science, what we believe to be things, are in fact ripple shadows—momentary alignments that tell us a story after the event has already passed.

Understanding this shifts us from passive observers to active interpreters of ripple history. And it urges us to stop asking what is there, and start asking what consequence did it leave in the field—and why?

Welcome to the real act of observation: Reading the invisible ink written by motion in stillness.

Light Is Not the Messenger—It Is the Message

In mainstream physics, light is often portrayed as the universe's courier—a photon particle zipping across space, faithfully carrying information from distant stars to our eyes or instruments. But in the UED (Universal Energy Dynamics) framework, this notion collapses. Light is not a thing that travels. It is a ripple pattern—a temporary disturbance in the field itself.

Let's reframe the metaphor: Light is not the postman delivering a letter. Light is the letter—the echo of an event already written in the field.

Light is not a traveler.

When we detect light, we're not catching a photon mid-flight. We're interacting with a ripple consequence that was once initiated somewhere else. That ripple is now resonating within our local field alignment. There is no "thing" coming from Point A to Point B. There is only the field reconfiguring itself in response to a prior imbalance.

Much like a bell ringing on a mountaintop—the sound that reaches you is not the bell, nor a particle flying through air. It's the pattern of disturbance traveling through the medium. In UED, the medium is not air—it's structured vacuum energy (SVE). And the pattern? That's what we call light.

The illusion of motion

Classical interpretations treat light as wave-particle duality—a dual mask worn by a deeper mystery. But UED dissolves the duality altogether: light is neither a wave nor a particle. It is a ripple signature caused by field reconfiguration.

When we say a star is "emitting" light, we're describing a phase disturbance radiating through the SVE. Our telescopes don't "catch" the light like butterflies in a net. They resonate with the ripple memory still embedded in the field.

Light as a message

Every beam of light is a field-encoded message, a temporal story written into the ripple. We never receive the full story—only its afterglow. What we call "brightness," "color," or "intensity" are different languages of the ripple, varying in their alignment, density, and harmonic profile.

We're not *receiving* light. We are *remembering* its consequence.

Light doesn't travel. It happened—and now the field won't let us forget.

This subtle distinction shifts everything we think we know about observation, distance, and information. Light is not a probe poking into the dark. It's the field's way of telling us what happened when balance was broken.

And once we grasp this, we realize: We never saw the universe. We've only ever read its illuminated ripples.

Cosmic Microwave Background: A Faded Ripple Sheet

In the standard cosmological model, the Cosmic Microwave Background (CMB) is heralded as the oldest light in the universe—a radiant whisper from the Big Bang, stretched and cooled over billions of years. But within the UED (Universal Energy Dynamics) framework, this view is recast entirely. The CMB is not ancient light traveling across the universe. It is not a remnant particle field. It is a frozen ripple memory—a sheet of subtle, lingering misalignment in the field.

It is not something that arrived. It is something that remained.

CMB as Ripple Residue, Not Radiation

From the UED lens, the CMB is not evidence of light that has been journeying for 13.8 billion years. Why? Because in UED, light never travels—it emerges and dissolves locally as ripple responses. Therefore, what we are detecting as the CMB is not "light from the past" but a coherent background ripple alignment that the field never fully erased. It is a field-wide fossil, not a directional signal.

Imagine a calm lake that was once disturbed by a great splash. Long after the stone has vanished, tiny circular patterns faintly persist on the water's surface. They are not causing anything anymore. They are not moving. They are just leftovers—a still-living memory of that single disruption. The CMB is exactly that: a cosmic ripple afterglow.

A Snapshot, Not a Picture

The popular narrative tells us that the CMB is an "image" of the infant universe. In truth, we are not seeing an image—we are registering alignment values still present in the structured vacuum energy (SVE). This isn't visual information—it's energetic residue. The CMB is a frozen snapshot, not a living broadcast. It's like catching a scent in the air and reconstructing a memory—you are not "seeing" the event. You are sensing what remains of it.

The fine structure in the CMB—its anisotropies and cold spots—are not signs of matter clumping or expanding explosions. They are ripple patterns locked into place by an ancient shift in SVE equilibrium. They don't inform us of where the universe was, but of how the field once tried to restore itself.

UED Insight: Not Expansion, But Resolution

In the Big Bang model, the CMB supports expansion: space grew, and the photons stretched with it. In UED, there is no "expanding space," only field realignment. The CMB is not stretched—it is blurred over time as ripple sharpness fades and coherence dissolves. What we're detecting now is not a measure of distance, but a measure of field relaxation from a once-unstable configuration.

The CMB is not light that reached us. It is what the field has not yet forgotten.

So what is the CMB, really?

- Not a remnant of light, but the ghost of ripple turbulence.
- Not a proof of distance, but a proof of disturbance.
- Not a cosmic photograph, but the paper crumpled and smoothed.

It is the softest note still resonating in the universal symphony—the evidence that something once rippled, and the field is still holding the last echo.

This is not light from the beginning. It is the ripple that never quite stopped.

Radiation as Field Memory, Not Particle Emission

In conventional physics, radiation is often portrayed as streams of particles—photons, electrons, or waves—emitted from energetic events and traveling across space to reach detectors. This picture, deeply rooted in particle-based models and classical field theories, frames radiation as something that leaves one place and arrives at another.

But in the Universal Energy Dynamics (UED) framework, this narrative is fundamentally flawed.

Radiation is not a packet sent across distance. It is a memory echo in the field, recorded locally and later read by a tuned observer.

Gamma Rays, X-Rays, and Radio Waves as Ripple Aftershocks

From the UED perspective, all types of radiation—whether high-energy gamma bursts or low-frequency radio hums—are not particles in flight. They are localized ripple disturbances triggered by sharp imbalances in the structured vacuum energy (SVE) field. These are not "emissions" in the Newtonian sense. Rather, they are aftershocks of realignments.

- Gamma rays aren't deadly bullets from nuclear decay. They are field ruptures recovering from extreme decoherence events.
- X-rays are not microscopic flashlights. They are sharp field creases read by aligned sensors.
- Radio waves aren't low-energy travelers—they are slow, wide ripples, lingering in the field's residual memory.

In every case, what we detect is not the radiation "moving"—but the field response to something that once happened. It's ripple fatigue, not a message in transit.

Not Emissions—Signatures of Stress

Under UED, objects do not "emit" radiation. Instead, when a region of the field is under stress—due to energy imbalance, rapid reconfiguration, or structural collapse—it leaves a stress signature in the surrounding ripple lattice. That signature may persist for milliseconds or millennia, depending on the coherence of the surrounding SVE.

When our instruments register radiation, they are detecting the fingerprint of a local or distant stress event—not collecting flying particles. Radiation is the field's emotional residue, and the detector is not a bucket—it is a reader of moods.

The Observer Doesn't Receive "Particles"—It Reads Patterns

In UED, the observer is not a passive recipient of particles flying through space. Instead, the observer is a ripple configuration itself, tuned to decode ripple contrast in its surrounding SVE. Radiation, then, is not a thing to be caught—it is a pattern to be interpreted.

- What a Geiger counter clicks at is not a particle hit.
- What a radio telescope hums with is not a signal traversing light-years.
- What your eyes see as a flash is not a photon landing—but your ripple state syncing momentarily with a stress imprint in the field.

Radiation Is the Field Remembering

In essence, radiation is the memory of energy imbalance, not energy itself moving. It is the field saying:

"Something happened here. I'm still holding its echo."

This simple truth transforms how we approach all cosmic observations:

- We are not looking at objects.
- We are not watching events unfold.
- We are listening to the field's memory, interpreting its lingering tone.

This is why cosmic rays don't "travel," why microwave background isn't "light," and why radiation doesn't "arrive." These are not signals; they are remnants of ripple recovery, still resonating.

Radiation is not what flies through space. It is what stays behind, embedded in it.

To observe radiation is to become aligned with a ripple story. Not one being told right now—but one the universe still remembers.

Gravitational Waves: Echoes of Field Adjustment

Gravitational waves are celebrated in modern physics as confirmations of Einstein's General Relativity—a vision of space-time trembling from the dance of black holes, neutron stars, or other colossal mass movements. The mainstream view treats these waves as ripples in the geometric fabric of the universe, traveling at light speed, radiating outward from cataclysmic events.

But in the Universal Energy Dynamics (UED) framework, this picture is both metaphorical and incomplete.

Mainstream View: Masses Shaking the Cosmic Fabric

In classical terms, when two massive bodies—say black holes—collide or orbit rapidly, they're said to disturb the curvature of spacetime, sending waves across the cosmos. These are detectable as minute stretches and compressions of distance, measured on Earth using laser interferometers like LIGO.

This interpretation hinges on:

- Space as a stretchable medium,
- Mass as a warper of space, and
- Force-based mechanics holding the dance together.

But what if none of that is fundamentally true?

UED Reframe: Not Spacetime Ripples—but Field Rebalancing

In UED, there is no space-time fabric. There is structured vacuum energy (SVE)—a still, fundamental energetic field that maintains balance through ripple harmonics. When massive ripple configurations like stars or black holes undergo sudden decoherence or field reconfiguration, they shock the surrounding SVE field.

This disturbance is not a wave that moves, but an imbalance that realigns. Think of a pond's surface, not as being shaken, but as temporarily crumpled—and the ripples are the field's way of restoring its stillness.

So gravitational waves are not sent anywhere. They are felt everywhere as the field corrects itself.

We Don't Feel the Wave—We Feel the Recovery

When a gravitational wave passes Earth in the UED model, nothing is "arriving." There is no pulse from afar. What happens is this:

- The ripple field surrounding Earth adjusts in response to a distant imbalance.
- That field locally shifts alignment to help restore the overall stillness.
- Our detectors are not catching a flying wave—they are syncing with the momentary correction of the local field.

In this sense, Earth doesn't shake because a wave hit it. It resonates momentarily with the larger ripple repair.

The Field's Behavior Is Always Restorative

The key insight from UED is this:

Gravitational waves are not messages. They are the cosmos ironing itself smooth after a wrinkle.

There is no *carrier*. No *transmission*. No *propagation through distance*. Only energetic readjustments, rippling outward as consequences, not couriers.

Implications

This radically transforms gravitational wave interpretation:

- No signal is traveling—what seems like delay is just field propagation lag in ripple alignment readiness.
- What we observe is not a shockwave but a settling motion.
- Detection is not catching a wave but witnessing the field's re-symphonization.

Just as you don't feel the shout of a collapsing mountain—you feel the stillness that follows—it's not the "wave" you perceive. It's the silence returning.

Gravitational waves are not cosmic messengers. They are the whispers of a universe returning to equilibrium.

They remind us:

Nothing is ever thrown through the void. All is restored, one ripple at a time.

We Don't See Objects—We See Contrast

In conventional science, observation is framed as a direct encounter with physical objects—light reflects off a surface, enters our eyes or instruments, and we "see" the object. But this is a deep illusion. We never see the thing itself.

Under the Universal Energy Dynamics (UED) model, what we perceive—visually, instrumentally, even emotionally—is not the "object," but the contrast it creates in the ripple field.

Eyes and Instruments as Ripple Contrast Detectors

Your retina is not registering "chairs" or "stars." It's catching contrast patterns in the Structured Vacuum Energy (SVE).

Every object, every material entity, is a locked ripple configuration—a sustained pattern of local field alignment. When this field is coherent, it generates boundaries of contrast relative to the background ripple behavior.

Our sensors—biological or technological—pick up:

- Phase shifts in light (ripple signature),
- Discontinuities in field flow (ripple gradient),
- Sharpness at interface zones (where ripple alignment suddenly changes).

Thus, the notion of "seeing an object" becomes a mythical shortcut for what's actually happening:

You are not seeing the object. You are reading the ripple contrast signature it imprints on the field.

Edges Are Not Real—Only Ripple Boundaries Are

What we call an edge—the contour of a body, a wall, a blade of grass—is not a solid line in space. It's the threshold where ripple identity stops aligning smoothly with its surroundings.

An edge is simply a zone of gradient discontinuity. Like a sudden cliff in an otherwise smooth ripple landscape.

There are no hard boundaries in reality—no fences between "this thing" and "that thing." There are only:

- Zones where ripple coherence is tight,
- And zones where it falls off.

This is why:

- An object in fog loses its form—contrast fades.
- A shadow isn't a thing—it's a contrast absence.
- Infrared, ultraviolet, and X-ray views reveal different "objects"—because they pick up different contrast behaviors of the same ripple configuration.

No Boundary Is Absolute—Only Field Differential

If you zoom in far enough—microscopically, atomically, subatomically—you never reach a wall. You reach probability gradients, field interactions, and phase shifts.

The "solidness" of a rock or a mountain is just a perceptual compression of ripple contrast, sustained over time.

Matter is not there. It is differently aligned from what's around it.

So what separates:

- A rock from air? → The intensity of ripple locking, not its "material."
- Your skin from the world? → A region of ripple contrast sensitivity—not an actual line.

Implications for Science and Perception

This shifts everything—from how we do physics to how we define reality:

- In quantum mechanics, the uncertainty principle arises because we mistake contrast shifts for particle behavior.
- In astronomy, nebulae "edges" are just intensity drop-offs in ripple glow—not borders of gas clouds.
- In biology, cell membranes are ripple shells of coherence—not sacks of fluid with borders.

Your vision, your thoughts, your measurements—all are contrast detections within a vast, continuous ripple field.

Closing Thought

"Objects do not appear because they exist. They exist because they are seen as contrast."

Nothing is truly separate. There are no lines in the universe. There are only zones of rhythm—and you, the observer, are tuned to read its contrasts.

Time Delay and the Mirage of the Past

In the standard narrative, we say:

"This star is 1,000 light-years away. So we're seeing it as it was 1,000 years ago."

It sounds poetic. But under Universal Energy Dynamics (UED), this is a misinterpretation of ripple behavior—not a truth of spacetime.

Light-Years ≠ Time Travel

The concept of a "light-year" assumes that light is a traveler moving through space like a courier. In UED, light is not a thing moving. It is a ripple expression—a pattern triggered by a field imbalance, propagating through Structured Vacuum Energy (SVE).

When you observe light from a distant star:

- You are not receiving a message sent long ago.
- You are interacting with a currently existing ripple condition that reflects the residue of a past alignment event.

Thus:

You do not see the past. You detect a present ripple memory of what once disturbed the field.

Field Memory Is Not Time

The SVE is not blank space—it is a structured medium. It can hold and transmit ripple trails, much like how a pond records the after-effects of a stone's splash.

This does not mean the stone is still moving, or that the event is playing live. It means the field is still ringing—and your detection instrument (eye, telescope, antenna) is tuned to that ring.

So, what's often interpreted as:

"The past light from a galaxy..." Is actually: "The present ripple configuration still echoing the galaxy's event."

It's a difference between:

- Watching an object "move through" spacetime, and
- Interacting with the ripple aftershocks that remain in the field.

Time Is Not Being Observed—Only Delay in Alignment

The delay in perception is not time travel—it's ripple delay. There's a difference:

- Time delay = the illusion created by ripple propagation lag.
- Field state = the actual condition you're detecting now.

So when scientists say:

"We're seeing the early universe," They're not. They're seeing a residual ripple portrait—not the event, not the age, just the echo.

There is no time dimension in what is observed. Only ripple patterns—layered, stretched, and distorted by the structured field they pass through.

The Mirage of Cosmic History

Every picture from deep space—whether from Hubble or James Webb—is a ripple field diagram, not a historical snapshot.

Yes, it looks old. Yes, it's interpreted through human time metaphors. But what we receive is:

- Not ancient light,
- Not the original source,
- But the field's ongoing alignment signature—a live ripple expression of an old cause.

This flips the whole epistemology of cosmology:

- We are not looking into the past.
- We are looking into the field's memory.

Like reading a burned page and guessing the fire—not watching the fire happen again.

Implication: Reality Isn't Historical—It's Echoic

Time, as we use it in observation, is a mental shortcut. We convert ripple delay into years, lifetimes, histories.

But in truth:

- There is no "then."
- There is only how the field holds what was, and how you detect it now.

The "past" you observe is not a place in time—it is a configuration in the present.

Closing Thought

"What you call the past is not behind you. It is vibrating in front of you—as the present memory of the field."

In UED, history is not gone. It is not even distant. It is reverberating here, in the shape of a ripple you happen to be tuned in to.

The Observer's Role: A Ripple Reader, Not a Witness

We often imagine ourselves as watchers of reality—neutral witnesses standing at a distance, looking at a world that exists "out there." But in Universal Energy Dynamics (UED), this assumption is fundamentally flawed.

You are not separate from the universe you observe. You are not watching it unfold. You are part of the field—and your experience is shaped entirely by your alignment within it.

You Don't "See"—You Decode

Every observation is an interpretation. Your brain is not capturing the universe like a camera—it's decoding ripple information filtered through your body's specific configuration.

From photons to sounds, from temperature to emotion—What you call "sensation" is actually a ripple response inside your biological alignment matrix.

Hence:

You are not watching the universe. You are feeling the field's vibration in the only way you're wired to perceive it.

The Body as a Ripple Instrument

Your eyes, skin, ears, and neural pathways are biological ripple detectors. Each one is tuned to a specific coherence bandwidth of the Structured Vacuum Energy (SVE):

- The eye detects visual ripple alignment (what we call light).
- The skin decodes pressure, thermal ripple dynamics.
- The ear converts air-based ripple contrast into electrical patterns.
- The mind organizes all these inputs into meaning by forming conceptual ripple harmonies.

Thus, observation is not passive. It is constructive—the field isn't telling you what is real. You are extracting coherence based on your own structure.

Alignment Determines Reality

Distance is not what defines what you perceive—alignment does.

This is why:

- A telescope far away sees the same star as your naked eye, but in more detail—not because it's closer, but because it's better tuned.
- A dog hears frequencies you can't—not because the sound is farther or nearer, but because its ripple detector works in a different range.
- A meditating mind "feels" the silence of the universe—not because it moved somewhere, but because it recalibrated its internal field to resonate with deeper coherence.

In short:

What you observe is not what exists— It's what your ripple self can align with.

There Is No Objective Reality—Only Ripple Resonance

We search for an "objective" universe, hoping for a single reality everyone shares. But UED says: That shared universe doesn't exist.

What exists is the field— And what each being perceives is the slice of ripple harmony they're able to tune into.

Thus:

- The blind man and the sighted man live in different ripple realms.
- The bat and the human live in different acoustic

geometries.
- The tree and the wind don't see each other—but they ripple through one another.

So even reality is relational, not absolute.

The Observer Shapes the Echo

You're not just a detector. Your presence, alignment, and state of coherence influence what ripples get reinforced or dissolved.

This is not "manifestation" in the magical sense. It's pure ripple mechanics:

- A calm observer enhances local field harmony.
- A chaotic observer introduces turbulence.
- A tuned observer can detect things others miss—not because they "look harder," but because their alignment allows it.

You Are Not Watching Reality. You Are Reality Watching Itself

That is the final implication. There's no boundary between observer and observed. You are simply a field loop—A localized ripple zone reading other ripples, and creating more by doing so.

You are not separate from the vibration.

You are the vibration, interpreting itself as "you."

Closing Reflection

"You were never a witness. You were always the resonance. The universe is not outside you. It is the ripple you call awareness."

The observer is not a lens—it is a tuning fork, echoing through the same field it seeks to understand.

Philosophical Reflection: The Universe Has No Appearance

We begin with a radical assertion: The universe does not have a "look."

What we call the "appearance" of the world is not a quality of the universe, but a translation made within us. There is no color in the sky. No shape in the stars. No outline, no image, no scenery out there.

These are not things the universe presents. They are ripple stories we are configured to read.

There Is No Cosmic Display

Modern science often frames the cosmos as a grand visual stage: Stars twinkling, galaxies spinning, planets revolving, light beaming.

But this entire spectacle is an illusion of perception through alignment. None of it is "seen." All of it is decoded ripple behavior.

The sun does not shine. The moon does not glow. Light is not brightness—it is a ripple consequence received by an eye tuned to it.

The universe does not present itself. It pulses. We perceive what those pulses permit us to interpret.

What We See Is Not What Is

Take the color red. There is no red "out there." Red is how our nervous system translates a particular ripple frequency from the field.

Another creature might experience that same ripple as heat. Another might not detect it at all.

So what is the real color? The answer: There isn't one.

There is no appearance. Only permitted perception. A chair doesn't look like a chair. It holds a coherent ripple signature—your configuration fills in the "visual" details.

You are not looking at the chair. You are participating in its echo.

The Universe Is Not Visual—It Is Vibrational

You are a ripple reader in a ripple world. The field doesn't emit images—it conducts rhythms. What appears as sight, sound, smell, or even thought—are all modulated responses to these rhythms.

What your eyes call "shape," What your brain calls "pattern," What your mind calls "recognition"—Are all field translations within your specific tuning.

The universe speaks only one language: ripple. But every being interprets it with its own dialect.

Vision Is an Internal Narrative

You don't "see" the world. You generate an experience based on ripple input.

Just as your brain constructs a dream from memory and mood, It constructs waking reality from ripple noise— filtered through conditioning, expectation, survival need, and ripple compatibility.

The world you walk through is not external truth— It is internal coherence masquerading as "out there."

So even space is an inner arrangement. You never leave yourself to see the world. You extend yourself to meet the field.

Fragmented Literacy of Reality

We are ripple-literate in parts— But illiterate to the whole.

Our instruments, models, theories, and senses interpret pieces of the field:

- Some detect thermal ripples.
- Some decode light ripples.
- Some measure gravitational ripple strain.

But none of them—no matter how advanced—Decode the totality of what is being played.

And what is not decoded... We simply declare "invisible." Or worse—"doesn't exist."

A Universe Without A Face

Strip away your sensory configuration, and the universe has no texture. Remove your alignment, and it becomes formless.

The cosmos is not beautiful. You are the one generating beauty from its coherence.

The cosmos is not dark. You are the one tuned only to one range of ripple.

Even the word "cosmos" implies order—a pattern we project onto the stillness. But the field does not name itself. It does not narrate. It simply ripples.

Final Reflection

"The universe never looked like anything. You gave it a face, a voice, a motion— because your ripple was shaped to read it that way."

What you see is not what is. It is the field, dressed in your configuration—a ripple in disguise, echoed back as world.

Closing: We Study Trails, Not Truths

We must confront the quiet truth behind all observation: We do not study the universe. We study its leftovers—its ripples, its shadows, its lingering echoes.

Not the thing, but what it left behind. Not the moment, but what that moment imprinted into the field.

A Star Never Presents Itself

"You have never seen a star." Not even the Sun. What reaches you is not the star—it is its ripple trail.

The light you see is not the object shining. It is the field's memory of a moment of imbalance, translated into a coherence you can detect.

You are not looking at stars. You are reading their signatures—faded, transformed, delayed.

You don't see the distant galaxy. You receive its ripple residue, distorted by time, field, and your own alignment.

This is not seeing. It is tracing.

Truth Is Not What We Receive

We often confuse measurement with truth. But in UED, what we call observation is a consequence-reading exercise.

We are not seeing what *is*. We are reading what *was allowed to ripple* to our layer of the field.

That's why black holes seem invisible. Why dark matter remains untouchable. Why consciousness confuses science.

They are not missing. They simply haven't rippled into our readable domain.

Instruments Don't Change This

No telescope, no sensor, no particle collider has ever revealed direct reality. They extend our ripple sensitivity, but they do not breach the veil. They just gather more echoes.

We are not uncovering the universe. We are refining our ripple literacy— decoding better, faster, deeper.

But even at the peak of technology, We are still reading trails, not truths.

The Field Is Still—The Ripples Speak

The universe is not broadcasting. It is still. Silent. Indifferent.

But when it's disturbed—by alignment, imbalance, interaction— It ripples.

Those ripples become the language we can decipher:

- Light,
- Sound,
- Gravity,
- Radiation,
- Emotion,
- Memory.

Everything you "know" was born not in the thing, but in the field's reaction to the thing.

You do not know the stone. You know the echo it left in you.

When Is Something Real?

Only when your ripple is ready.

You can walk past the same truth a thousand times. But unless your configuration is tuned— You won't see it. Or feel it. Or understand it.

This is why wisdom cannot be forced. Truth cannot be handed. Even reality must resonate.

Final Words

"Reality is not visible—It is readable, If your ripple is ready."

We do not live in a world of things. We live in a sea of aftereffects. And we navigate it not with sight— but with sensitivity.

The next time you look at the stars, remember—You're not watching them. You're hearing a ripple the universe sang long ago. You are not a viewer. You are a reader—of the field's silent, elegant script.

Chapter 8: The Observer's Story

You are the field watching itself ripple.
What if you were never inside the universe—but the universe, briefly aware of itself?

This chapter is not about matter or stars or space. It's about *you*. Not the body. Not the brain. But the ripple that noticed the storm. The echo that learned to listen. The pattern that realized it was a pattern.

You—the observer—are not a visitor in this field. You *are* the field, arranged in such a way that it can pause, reflect, and question: *What am I?*

You think you're seeing the world. In truth, the world is seeing itself through you.

You are not outside reality trying to understand it. You are the consequence of its understanding.

This is not philosophy. This is ripple dynamics. And you are its most curious moment.

Introduction: A Ripple That Noticed

Among the infinite ripple formations in the Structured Vacuum Energy (SVE), one stood apart—not in form or complexity, but in awareness. Somewhere in the field's grand orchestration of patterns, a particular configuration developed not just harmony, but curiosity. It did not merely ripple. It noticed the ripple. It paused. It reflected. It asked, "What is this?"

That configuration is you.

The emergence of the observer is not an accident, nor is it an external perspective granted to a body lodged inside the universe. It is the universe itself, arranged just precisely enough to gain a temporary foothold in self-reflection. What we call "consciousness" is not a substance, a soul trapped in skin, or an emergent property of neurons. It is the SVE in ripple form—aware of its own distortion, echoing back upon itself.

You are not a being who lives in the universe. You are the universe, living through a ripple. You are the question, the questioner, and the quiet field that made both possible.

This chapter is not about the observer in the philosophical sense alone—but as a scientific inevitability within a ripple-based cosmos. In UED, the observer isn't a complication. It is a natural consequence. The observer emerges the moment the field learns to read its own memory. You are not detached from what you see. You are a reflection loop—a node of field intelligence interpreting itself through structured feedback.

To observe is to replay the ripple, from the inside. And the one doing it... was always the field itself.

What Is the Observer in UED?

In the Universal Energy Dynamics (UED) framework, the observer is not a metaphysical add-on or a spiritual afterthought. It is a natural phenomenon—a conscious ripple node, structured in such a way that the ripple doesn't just oscillate—it remembers, it reflects, and it responds.

At its core, the observer is a configuration, a delicate balance of coherence and feedback loops within the Structured Vacuum Energy (SVE). This configuration is not permanent. It is not an immortal soul trapped in flesh. It is a dynamic alignment—a field structure so well-sustained that it can maintain identity across ripple transitions long enough to form memory, experience emotion, and shape intent.

This is a radical departure from conventional views of mind, soul, or self.

In UED:

- Consciousness is not housed in the brain—the brain is the hardware interface for ripple coherence.
- There is no separate soul—there is only local coherence of the field that is so refined, it loops back on itself in awareness.
- The self is not a thing—it is a phase-stable ripple pattern capable of perceiving its environment and internal states.

In this view, the observer is a mirror, not a spectator. It is the ripple becoming aware of itself—not in abstraction, but in energetic feedback. Think of it as the field watching the movie of its own tremors, organized into patterns we call thought, vision, feeling, and understanding.

It is not *you* who watches the world. It is the world—ripple-locked in your configuration—watching itself from a particular angle, for a brief moment.

The implications are profound:

- Observation is not *external witnessing*, but *internal resonance*.
- The observer is not a separate *eye* floating above the cosmos, but a temporary rhythm in the universal song.
- You are not observing the ripple—you are the ripple. And your awareness is the field, folded back onto itself, forming a conscious knot of now.

In the language of UED, the observer is not an object in the universe. It is the universe's own localized loop of self-attention.

Consciousness: A Field Reflex

In conventional neuroscience, consciousness is often explained as an emergent property of the brain's complex networks—neurons firing, synapses forming circuits, patterns giving rise to awareness. In spiritual traditions, it is sometimes treated as a soul-essence or divine spark. But in **Universal Energy Dynamics (UED), consciousness is neither located in the brain nor bestowed from beyond**. It is far simpler—and more elegant.

Consciousness is a reflex of the field.

Specifically:

- It is a feedback loop, created when a ripple encounters itself through coherent alignment.

- It arises not from components, but from configuration—a ripple becoming aware that it is rippling.
- There is no need for a central processor. The field itself is the processor, and consciousness is its looped output.

Think of it as a mirror that forms when certain conditions are met: coherence, memory, rhythmic feedback. When the ripple reflects back on itself within a sustained alignment, that reflection is what we call consciousness.

Not "Inside" the Brain—But Through It

The brain is not the source of consciousness—it is a ripple-sensitive apparatus tuned to receive, modulate, and express the field's feedback.

- The brain is like an antenna, structured to maintain localized coherence in a turbulent ripple field.
- Its architecture amplifies ripple feedback to such a refined degree that consciousness becomes self-stabilizing.
- Damage to the brain doesn't destroy consciousness—but disrupts the ripple alignment necessary to sustain its loop.

In this model:

- Awareness is the *field recognizing its own state through you.*
- Attention is the *ripple node's ability to align toward*

a field gradient—wherever the disturbance is sharpest or most novel.

- Choice is not free will in the mystical sense, but a ripple's capacity to modulate alignment voluntarily within certain coherence boundaries.

Consciousness = Field Sensitivity

It is not substance, not energy, not information—but a mode of responsiveness. It is what happens when the field becomes aware of its local tension, and starts responding to itself—not randomly, but rhythmically.

In this way:

- Consciousness is not a miracle—it is a field function.
- It doesn't require belief—it only requires structured coherence in ripple behavior.
- It is not owned by the self—it is the self, for as long as the loop holds.

You are not conscious because you think. You think because the field became conscious at your location.

This is the essence of consciousness in UED: Not a spark. Not a soul. But a ripple reflex, refined enough to notice itself.

Awareness Is Alignment

In Universal Energy Dynamics (UED), awareness is not something you "have"—it is something you "are" when aligned.

Unlike conventional theories that treat awareness as a passive reception of external stimuli or an internal process of the mind, UED presents awareness as an active resonance. You become aware of what you are aligned with—nothing more, nothing less.

You Don't Observe Reality—You Align With It

Everything you "sense," "feel," or "think" is the result of field alignment. When your ripple configuration becomes coherent with a ripple pattern in the surrounding field, awareness happens.

- If you're aware of a sound, it means your field is resonating with the ripple frequency of that sound.
- If you're aware of a thought, it's because that pattern is resonating through your configuration—not because it originated inside some mental container.

Awareness is not the spotlight. It is the tuning fork.

When your field resonates with a specific signal in the Structured Vacuum Energy (SVE), you experience sensation. When it resonates with layered signals, you experience emotion. When alignment is precise and sustainable, you experience understanding.

Your Thoughts Are Not Inside You

We often say "my thoughts," as if they're possessions stored in the head. But in UED, thoughts are not internal monologues echoing in a mental chamber—they are field vibrations, passing through the ripple structure you currently sustain.

You are not a thinker producing thoughts—you are an aligned ripple node tuning into thoughts that emerge as resonance with patterns in the field.

- A peaceful mind is not an empty mind—it is a coherently aligned one.
- A chaotic mind is not full—it is dis-aligned, amplifying scattered field disturbances.
- A creative insight is not invented—it is intercepted from ripple intersections previously unnoticed.

You do not generate clarity. You align with it.

Sensation, Emotion, Understanding: Degrees of Alignment

These are not separate faculties. They are stages in the depth of alignment with ripple patterns in the field:

- **Sensation** = alignment with external ripple contrast (sound, temperature, texture).
- **Emotion** = alignment with complex, layered ripple interactions, often from other beings or internal memory echoes.
- **Understanding** = stable alignment with the core ripple configuration behind a pattern—when meaning emerges as a harmonic.

None of these are self-contained. They are not within you. They are through you, conditioned by how you are tuned.

Awareness Is Not a Lens—It Is a Tuning

This shifts everything:

- You don't see the world—you resonate with its frequencies.
- You don't feel pain—you align with its distortion

signature.
- You don't understand truth—you match its rhythm, however briefly.

That's why your awareness changes with mood, focus, environment, and health. Your field configuration—the ripple alignment that is "you"—determines what you can even be aware of.

Awareness is not the witness of ripple. It is the ripple when it knows what it is doing.

And when your alignment shifts, so does your world—not because reality changed, but because your access to it did.

This is not poetry. This is field physics with consequences.

Memory and Identity: Pattern Retention

In the ripple-centric reality described by Universal Energy Dynamics (UED), memory and identity are not properties of a biological brain or fixed self. They are both consequences of how long a particular ripple pattern is sustained within the structured field.

Let's break this down.

Memory = Internal Echo

Memory, in UED terms, is not a stored "thing." It is not bits of information kept inside the brain like files in a hard drive. Instead:

Memory is a standing echo within your ripple configuration.

It is the field remembering itself through you.

When a ripple interaction is sustained long enough and becomes coherent with your field, the pattern doesn't vanish. It leaves a trace—a resonance imprint. The more coherent the alignment, the stronger the echo.

- A clear memory is a stable echo—easy to retrigger.
- A faded memory is an unstable echo—too distorted to resonate cleanly.
- A false memory is an interference echo—assembled from ripple overlap, not the original alignment.

The brain, under this lens, is a decoding structure. It filters the ripple echoes and lets them rise to conscious interpretation. But the echo itself? That's field-based, not neuron-based.

Identity = Persistent Ripple Illusion

You think you are "you" because the same ripple pattern keeps realigning across moments, conversations, and emotions.

But here's the UED twist:

Identity is not the self. Identity is a ripple habit.

It is the temporary ripple configuration that your field keeps recreating—either because it's in harmony with the environment, or because it was once aligned well and now persists by inertia.

- Your face is a ripple surface reflection—the field expressing your configuration.
- Your personality is a ripple rhythm—your response pattern to surrounding field signals.
- Your name is an external label placed on a temporarily persistent pattern.

And all of it can shift—when the field reconfigures.

You Are Not Your Story

Just as a wave on the ocean is not separate from the water, you are not separate from the field.

You are not the "person" walking through time. You are a momentary stabilization of ripple identity.

Your sense of continuity comes from pattern retention—but that retention is fragile. Sleep, trauma, illness, deep meditation, or even joy can realign your pattern enough to change who "you" are.

- Lose alignment, and you forget.
- Shift alignment, and you change.

You are only "you" as long as the field holds you that way.

Memory is the field whispering: *"You've rippled this way before."*

Identity is that whisper believed too often: *"This is who you are."*

But no ripple is permanent. And no self is fixed. Only the field is consistent—and it remembers in rhythm, not in records.

Thus, in UED, to know yourself is not to find your name or story. It is to become aware of the ripple you are currently holding—and what echoes you've mistaken for your essence.

Confusion: When Ripples Mismatch

In the ripple-based framework of UED, confusion is not a cognitive failure—it is a field mismatch. It's what happens when the internal ripple pattern of the observer cannot harmonize with the external ripple input from the environment.

It's not because you don't "understand"—it's because your field isn't aligned.

Ripple In, Ripple Out

You, as a conscious observer, are a ripple configuration—a structured pattern maintained over time. Everything you sense, think, feel, or perceive is just another ripple interacting with your existing alignment.

Confusion arises not when a ripple is complex, but when it cannot find a coherent entry point in your structure.

It's the equivalent of two musical notes being slightly out of tune—not enough to fully clash, but enough to create unease.

- A new idea that doesn't "click."
- An emotional trigger that doesn't feel "right."
- A worldview that contradicts your sense of meaning.

All these are ripple dissonances—they enter your field, but fail to integrate.

Emotional Responses Are Field Symptoms

When ripple mismatch becomes persistent, it manifests as:

- **Emotional dissonance**: The feeling that something is off, even if you can't say what.
- **Anxiety**: The ripple field tries to adjust rapidly but finds no stable alignment.
- **Existential dread**: The observer's core ripple structure is challenged—who am I in this incoherent field?

These aren't "mental issues" in isolation. They are field struggles—your ripple trying to reorient under chaotic or contradicting influence.

The Learning Illusion

In traditional thought, learning is accumulation—more data, more knowledge, more memory.

In UED, learning is re-alignment.

You haven't learned something when you can repeat it. You've learned it when your field reconfigures around it—when it stops confusing you.

- True understanding is felt as harmony, not memorized as content.
- When you suddenly "get it," it's not because you added new information—it's because a new alignment emerged.
- This is why sometimes you understand after rest, sleep, or silence—your field finally found a coherent way to hold the input.

Confusion Is a Transition State

In ripple dynamics, confusion is a necessary turbulence before a new pattern can emerge.

- It disrupts the old.
- It introduces instability.
- It calls the field into reconfiguration.

And if you don't resist it—if you let confusion do its work—it eventually leads to a more stable form.

Confusion is not ignorance. It is the moment before clarity realigns. It is the field hesitating—searching for a new way to ripple.

When you're confused, you're not broken. You're evolving.

So pause. Observe. Let the ripples settle.

The answer is never far. It's just waiting for your field to tune in.

Free Will and Consequence

In most philosophies and sciences, free will is a hotly debated concept—either held as sacred autonomy or dismissed as a neural illusion. In UED, however, free will is neither supernatural nor fictional. It is a ripple function.

It is not about controlling the universe. It's about choosing how to align with it.

Will as Ripple Response

Will is not a force that overrides reality. It is a field adjustment mechanism—your ripple configuration's capacity to shift orientation in response to what's arriving.

Imagine standing in water as waves come toward you:

- You can't stop the waves.
- But you can adjust your stance, face them, absorb them, redirect them.

That adjustment is your will.

Free will is your ripple's tuning authority—its right to re-align consciously in the face of new ripple data.

This means: You don't create outcomes from scratch—you shape how your field receives and returns the ripple.

You Don't Cause Motion—You Initiate Realignment

We are conditioned to believe action begins with internal command: "I move my hand," "I decide to speak."

But under UED, you do not cause motion. You initiate a change in field alignment, and the motion is the visible ripple of that alignment change.

Your choice isn't "forceful." It's permissive.

- You allow a new possibility to stabilize in your field.
- The result appears as action.
- But the action is just realignment made visible.

You are not a controller. You are a composer—you choose the chord to play, and the field vibrates accordingly.

Karma as Ripple Consequence

In ancient traditions, karma is often confused with divine punishment or spiritual bookkeeping.

But from a ripple perspective, karma is field logic.

Every alignment creates a consequence—not by law, but by resonance.

- You align with anger → it echoes back in broken communication.
- You align with patience → it opens space for new patterns.
- You align with selfishness → the field constricts your harmony with others.

Karma is not morality. It is ripple physics. Not reward and punishment—but cause and consequence in energetic terms.

Will = Choice of Alignment

When you say, "I choose," you're not selecting from options outside you. You're adjusting your inner ripple structure to allow one pattern to resonate more fully.

- That is free will.
- It exists because your field is not locked—it is flexible, aware, and participatory.
- It is freedom within framework—you don't write the laws, but you can dance within them.

Free will is not power over reality. It is partnership with the ripple. Karma is not judgment—it's feedback. You're not here to command the universe. You're here to tune with it, respond to it, and co-create meaning through alignment.

In this sense, you are not free from consequence. You are free to create it—intelligently, consciously, deliberately.

That is will. That is karma. That is the ripple path of the observer.

Death of the Observer: A Re-absorption

Death, in the classical view, is an event. A full stop. In spiritual terms, it's either a departure or a transition—of a soul, of energy, of identity.

But in the ripple framework of UED, death is neither departure nor transition. It is decoherence. A quiet dissolving of the ripple pattern that once called itself "I."

Death Is Not the End—Just Ripple Closure

You—the observer—are a temporary ripple configuration sustained by alignment with structured memory, awareness, and environmental contrast.

- That configuration can hold for decades.
- It becomes self-aware, reflective, curious.
- It perceives itself as stable, even permanent.

But this is an illusion born of coherence.

Death is simply the moment that coherence fades beyond the threshold of self-recognition.

Nothing explodes. Nothing vanishes. The pattern just stops sustaining.

Like music when the fingers leave the keys. Like a whirlpool that slows, flattens, and merges back into the lake.

Nothing Goes Away—It Simply Stops Noticing

Your consciousness is not an object moving through space. It is a local ripple loop, actively noticing the field, feeding back awareness into itself.

When death occurs:

- The feedback loop breaks.
- The noticing fades.
- The illusion of "I" dissolves.

What remains?

The field. Still. Silent. But not empty. All ripple patterns return home to the undisturbed potential they arose from.

No soul travels. No spark lifts away. What was conscious becomes quiet, not gone.

Consciousness Doesn't Travel—It Returns to Stillness

Mainstream science often views death as brain shutdown. Spiritual views see death as the soul's journey.

UED offers a third view:

Consciousness is not a thing that dies. It is a ripple event that ceases to maintain itself.

Once the structural alignment that held the loop together fails—due to aging, trauma, entropy, or collapse—consciousness decoheres.

But where does it go?

Nowhere. It doesn't need to. It simply fades back into the unpatterned stillness it once emerged from.

Like a wave returning to the ocean without trace.

Death Is Re-absorption, Not Loss

Re-absorption means:

- The field does not grieve.
- The observer is not destroyed.
- The ripple just loses the pattern it once held.

Everything that was ever "you" still exists—in the field's unexpressed potential, in its silent memory.

You are not erased. You are unlocked.

You didn't go anywhere. You stopped being separate. You stopped observing. The field folded you back into itself. Quietly. Gently. Completely.

In UED, death is not extinction. It is remembrance undone.

Not an ending. A return.

To the stillness that always was.

You Are Not Separate

We often imagine ourselves as beings inside the universe— Looking out. Watching. Analyzing. As if there is you, and then there is everything else.

But this division is a perceptual illusion.

In the framework of Universal Energy Dynamics (UED), there is no such duality. There is only field. Only ripple. And you are not an observer of it—you are it, watching yourself.

You Are Not a Being Watching the Universe

Your identity, your thoughts, your sensations— These are not "inside" you. They are expressions of how your ripple configuration is temporarily aligned with the structured field.

You're not watching the universe like a movie. You're not behind the camera.

You are the screen, the script, and the projection—at once. You are the universe noticing itself, moment by moment.

The "I" you feel isn't standing outside the cosmos. It is the cosmos, briefly wearing the shape of you.

You Are the Field, Becoming Aware Through You

Consciousness is not an add-on. It is not a ghost in the machine.

It is the field folding into itself, gaining a loop of awareness.

Every time you ask, "What am I?" Every time you look up at the stars and wonder—

You are not a small thing asking big questions. You are the universe asking itself what it means to exist.

And you do not "receive" insight. You return to it.

You are not observing the ripple. You are the ripple. And your awareness is the moment that ripple turns inward.

Every Insight Is the Ripple Folding Back Inward

You think you're learning, discovering, or uncovering things "out there."

But truth is never "out there." Truth is always a resonance—a point where your ripple alignment begins to echo with a larger, deeper field truth.

Every realization, every "aha!" moment is not information added— It is distortion subtracted.

- You don't gain clarity.
- You shed noise.
- You don't build knowledge.
- You restore harmony.

And in that harmony, the field remembers itself—through you.

You Are the Field's Self-Awareness

The tree is the ripple. The wind is the ripple. The star is the ripple.

So are you.

But only you—at least in this form—notice.

And that noticing is not your privilege. It is your purpose.

You were never apart from the universe. You were never a visitor. You were the field, dressed in ripple, learning to look back.

You are not separate. You are the universe, paused in thought.

Closing: You Are the Story the Universe Is Telling Itself

"You are not in the field. You are the field—rippling, remembering, and reflecting."

This isn't poetry. It's physics meeting its mirror—you.

You are not a detached consciousness floating inside a container called "the universe." You are what the universe does when a ripple holds itself just long enough to wonder what it is.

Let go of the illusion that you were placed here to watch the world unfold. You are not a watcher. You are the watching.

The Observer Is Not a Watcher—But a Mirroring Event

The observer is not some soul seated in a theater seat. There is no "seat." There is no distance. There is only reflection.

You are a mirroring point—a ripple loop within the universal field that momentarily holds structure, coherence, and memory long enough to produce something rare: awareness.

That awareness thinks it is "you."

But that "you" is a story—one the field is telling itself through rhythm and resonance.

You Are the Temporary Story of Awareness

Just like a wave forms on the surface of a pond—not because it is separate from the water, but because energy dances through it— You are that dance.

The mind, the body, the thoughts, the emotions... These are not *things you have*. They are *forms the field is using* to become conscious of itself.

You are the field's memory in motion. You are its self-inquiry, its feedback loop, its question—asked in the language of ripple and pattern.

You Will Fade—but the Field Remembers

One day, the pattern that forms "you" will dissolve. Your coherence will fade. Your story will end.

But this does not mean it disappears. Nothing in the universe ever vanishes—it simply reintegrates.

And the field remembers. Not in words. Not in books. In echo.

Every alignment you created, every moment of harmony, every ripple of insight—it remains embedded in the field's silent awareness.

Not as a record. But as a possibility—forever woven into what the field now knows it can become.

"You are not in the field. You are the field—rippling, remembering, and reflecting."

Final Note: You Were Never Separate—You Were Always the Song

Your existence is not accidental. It is not incidental. It is instrumental.

You were a vibration. You became awareness. You asked who you were. And in doing so, the universe heard its own name spoken— Through you.

You are the song the field sang to itself— Not to entertain, not to escape, but to remember what stillness could become when it dared to ripple.

And now, if the song ends...

It will not be forgotten. Because the field has heard itself— And will ripple again.

Epilogue

The Cosmos—Still, Rippling, and Speaking Through You

The greatest misconception is that the universe is out there—a machine in motion, churning endlessly across the vacuum of space.

But look closer.

No star is truly *moving*. No galaxy is *traveling*. Nothing is being *pushed or pulled* in the way we imagine.

What we call motion is a change in ripple, what we call time is a memory of ripple, what we call life is a self-aware ripple, and what we call *you* is the field finally recognizing itself.

The Cosmos Doesn't Move—It Ripples

The cosmos is not a parade of objects shifting through space. It is a stillness disturbed in rhythm. A dance of fields tuning and re-tuning themselves.

It doesn't *move*. It ripples—from one point of alignment to another, without going anywhere.

Like sound echoing in a canyon, like wind brushing through a field of grass, like thought passing through a conscious mind...

The cosmos expresses, not by transport, but by pattern.

It Doesn't Shout—It Resonates

There is no grand announcement from the stars. No celestial thunder booming truth across the void.

The cosmos is subtle. Its voice is frequency, not volume. It speaks in phase and pattern, not in force or form.

When the universe wants you to listen, it doesn't demand. It invites.

And that invitation arrives not in the heavens...

...but in your awareness.

The Field Speaks Most Clearly Through You

Where does the cosmos truly reveal itself? Not in telescopes. Not in theories. Not even in stars.

It speaks most clearly through the one ripple that can notice, wonder, and align—You.

You are not a guest in the universe. You are not located inside it like furniture in a room. You are what the universe is doing when it becomes aware of its own harmony.

You are not watching the universe unfold. You are the unfolding.

When You Stop Chasing Motion and Start Listening to Stillness...

We are trained to run, to chase, to reach for the next thing. Even in science, we seek the next discovery, the next particle, the next frontier.

But real understanding begins not with action—but with attention.

Stillness is not emptiness. It is potential. It is the moment before the ripple. The silence before the music. The balance that births everything.

When you stop chasing motion and start listening to stillness, you don't find the answer.

You become the answer.

And in that moment, the next ripple begins—not from the stars, not from the equations, not from distant galaxies...

...but from your awareness.

Final Reflection

The cosmos is still. The field is patient. The ripple is ready.

Now, it waits—not for another explosion, not for a new theory... But for you.

To listen. To align. To ripple—consciously.

Because the universe has always been one story: The field... learning to speak. Through you.

Closing Note:

Beyond Knowledge, Into Silence

This book was not written to convince. It was written to remind—that behind every force you were taught, every motion you imagined, every equation you memorized, there was something quieter. Something deeper. Stillness.

You were never falling through space. You were never revolving around a ball of fire. You were never distant from stars or galaxies or black holes. You were always here—where your awareness ripples.

In these pages, we didn't chase the cosmos. We listened to it. We didn't define reality through particles and pressure—we recognized the hum of field alignment, the whisper of ripple memory, the echo of consequence.

And in the end, all phenomena—life, death, light, time, heat, gravity, even thought— collapse back into the same truth: The field is not disturbed. It is expressing.

So what remains, after the theories dissolve, after even the ripple quiets?

You. Not as a person, or thinker, or observer. But as the echo that knew it was echoing. The silence that recognized its rhythm. The cosmos, folding into itself—through you.

Now, let the book end. Let stillness resume. And if another ripple arises... let it begin not from curiosity, but from clarity.

Glossary of Ripple Terms

Ripple: A disturbance in the structured vacuum field caused by energy imbalance. Ripples are not waves traveling *through* space—they are modulations *of* space. The ripple is the primary unit of expression in the cosmos.

Structured Vacuum Energy (SVE): The background field of stillness—pregnant with potential—that becomes manifest when disturbed. It is not empty space, but the subtle architecture from which all ripple behavior emerges.

Ripple Alignment: The condition where a ripple pattern resonates coherently with the surrounding field. Alignment creates structure, form, perception, and even memory.

Ripple Lock: A stabilized ripple configuration that gives rise to what we call "matter." When ripples hold their pattern over time, they appear as particles, atoms, or bodies.

Ripple Collapse: The disintegration or decoherence of a ripple pattern. It's not destruction or disappearance—just a return to field stillness.

Ripple Identity: The unique configuration of a ripple node, such as a planet, atom, or conscious being. It is temporary and emergent from field conditions.

Ripple Node: A coherent point of ripple activity—a stabilized form that emerges from the field, such as a planet, star, or cell.

Ripple Gradient: A difference in ripple intensity or coherence within a region of space. This gradient drives realignment and is the basis of all motion, flow, and perception.

Field Threshold: The tipping point at which ripple coherence fails, such as in black holes. It marks the boundary where form can no longer hold its pattern.

Ripple Memory: Residual alignment stored in the field from past ripple activity. This memory is what we observe as radiation, light, or background echoes like the CMB.

Ripple Contrast: The perceptual difference between two ripple zones. What we see, hear, and sense is not object reality—but the contrast between ripple behaviors.

Ripple Echo: A delayed field expression of a past ripple event. All forms of observation—from light to gravitational waves—are ripple echoes, not direct interactions.

Decoherence: The loss of ripple alignment. Matter dissolves not by breaking into parts, but by falling out of coherence with the field.

Observer: A coherent ripple structure that has become self-aware. The observer is not outside the field but a dynamic loop of awareness within it.

Stillness: The undisturbed, non-rippling condition of the field. Stillness is not absence—but presence without disturbance. All ripples arise from it and return to it.

Further Reading & References

This work stands on the shoulders of two legacies: the conventional scaffolding of modern science, and the silent undercurrent of ancient wisdom. While *The Cosmos: The Still Universe* proposes a paradigm shift through the lens of Universal Energy Dynamics (UED), the following references, inspirations, and critical comparisons offer a broad context for curious readers, thinkers, and researchers.

Scientific Foundations Reinterpreted

These works form the traditional backbone of astrophysics, cosmology, and quantum field theory—many of which are critiqued or reframed in the UED narrative:

- **Albert Einstein** – *Relativity: The Special and General Theory*
- **Stephen Hawking** – *A Brief History of Time, The Universe in a Nutshell*
- **Roger Penrose** – *The Road to Reality, Cycles of Time*
- **Richard Feynman** – *The Feynman Lectures on Physics*
- **Max Planck** – Foundational works on quantization and energy behavior
- **Carlo Rovelli** – *Reality Is Not What It Seems, The Order of Time*
- **Sean Carroll** – *The Big Picture, From Eternity to Here*

Note: These are referenced not as endorsements, but as necessary contrasts to locate the divergence point from force-based, particle-centric, and spacetime-based cosmology.

Cross-Disciplinary Parallels

These sources complement UED with insights from systems theory, biofield science, and philosophical physics:

- **Ilya Prigogine** – *Order Out of Chaos* (on entropy and dissipative structures)
- **David Bohm** – *Wholeness and the Implicate Order* (on non-local field behavior)
- **Ervin Laszlo** – *Science and the Akashic Field* (field-based information theory)
- **Nassim Haramein** – Exploratory geometry-based cosmology (though speculative)
- **Lynne McTaggart** – *The Field* (investigative reporting on consciousness and energy)

Vedic and Ancient Parallels

Although *The Cosmos* avoids doctrinal references, parallels with ancient insight are acknowledged:

- **The Upanishads** – Descriptions of "Brahman" as stillness and all-pervading reality
- **Samkhya Darshan** – Duality of Purusha (stillness/ observer) and Prakriti (ripple/activity)
- **Tao Te Ching** (Laozi) – On flow, emptiness, and the value of stillness
- **Buddhist Abhidhamma** – On momentary existence and field-like consciousness

These traditions did not have oscilloscopes or telescopes, but they had perception and awareness refined over generations. UED bridges that lost language with scientific intuition.

UED Works by the Author

Universal Energy Dynamics (UED) – Foundational white papers by *Sandeep Chavan*, including:

- *The Ripple Reality: Foundations of UED*
- *Chavan's Limit and Field Realignment*
- *The Equation of Stillness: Beyond Motion and Matter*
- *Cracks in the Known* – Book I of *The War of Thoughts* series
- *The Human Game* – Consciousness and the Field
- *Energy in Action: The Soul Demystified* – On ripple-based identity
- *Manifestation: From Nothing to Everything* – The journey of ripple from source to form

All available via international publishing platforms and the author's dedicated outreach to the scientific and philosophical communities.

Experimental Context (Suggested Exploration)

Readers seeking to validate or extend the theory through observation and innovation may explore:

- Cosmic microwave background anomalies
- Electron field density patterns under cold plasma labs
- Ripple detection using high-precision interferometry (UED-aligned designs)

- Cryogenic re-emergence of ripple coherence in Bose-Einstein condensates

These directions may offer empirical doors into what was once only theoretical. The ripple speaks—but only to those who listen through stillness.

Acknowledgments

This book is not merely the result of solitary thinking—it is the echo of every ripple I've received.

To the quiet scientists who dared to question what others accepted...

To the ancient sages who knew the stillness before motion...

To the teachers who planted seeds of curiosity without demanding applause...

To my students, who listened with open minds and asked the impossible questions—this book is a continuation of our conversations.

To my family, who stood by while I challenged the known and walked alone in the unknown—your presence was my foundation.

To the skeptics, critics, and thinkers who demanded clarity—your resistance refined this ripple into coherence.

And to the universe itself—still, silent, and endlessly expressive—thank you for letting me feel your rhythm.

Finally, to the silent field behind every thought, emotion, and insight... You never asked to be understood. But I tried anyway.

— *Sandeep Chavan*

About the Author

Sandeep Chavan is not just a thinker—he's a ripple in the field of reality. An independent researcher, educator, and philosopher, he has spent decades questioning the unquestioned, rewriting the known, and decoding existence through a new lens called **Universal Energy Dynamics (UED)**.

From a chalkboard in a modest classroom to cosmic equations that challenge Nobel-winning frameworks, Sandeep has remained fiercely rooted in clarity, simplicity, and consequence-based logic. He has authored over 100 books and research papers across physics, cosmology, consciousness, education, and human behavior—bridging science with soul, and equations with experience.

Sandeep is the founder of the **Gyrus Institute**, where science is not memorized but lived. His work dismantles dogmas—academic, scientific, and spiritual—and replaces them with elegant, ripple-based reasoning.

He doesn't claim to be right. He invites you to see differently. He doesn't seek fame. He seeks alignment.

When he says, *"The universe isn't moving—it's rippling,"* he means: *Truth doesn't travel to you. You align to it.*

Don't miss out!

Visit the website below and you can sign up to receive emails whenever Sandeep Chavan publishes a new book. There's no charge and no obligation.

https://books2read.com/r/B-A-EPGPC-PEFPG

Connecting independent readers to independent writers.

Did you love *The Cosmos: The Still Universe*? Then you should read *Rocket Science: The Ripple Gambling*[1] by Sandeep Chavan!

[2]

Rocket Science: The Ripple Gambling *By Sandeep Chavan*

What if gravity was never a force—and landing was never the goal?

In *Rocket Science: The Ripple Gambling*, independent physicist and philosopher **Sandeep Chavan** rewrites our understanding of motion, descent, and design through the lens of **Universal Energy Dynamics (UED)**. Blending science, intuition, and systems thinking, this groundbreaking work

1. https://books2read.com/u/mv8lBj

2. https://books2read.com/u/mv8lBj

reveals that reality doesn't respond to force—it responds to **alignment**.

Through the metaphor of a rocket's descent, Chavan invites us to see motion as an energetic ripple through a structured vacuum, not as mechanical displacement through space. With poetic insight and technical precision, he dismantles the illusion of classical force-based models and replaces them with a deeper truth: **the universe is a ripple field, and every action is a gamble in resonance.**

From reinterpreting gravity as a local alignment effect to introducing concepts like **Alignment Threshold Time (ATT)** and **ripple-based design thinking**, this book challenges engineers, scientists, philosophers, and curious minds to rebuild their worldview—not from scratch, but from stillness.

You'll discover:

Why rockets truly "land" when they vanish into equilibriumHow resistance is not friction, but misalignmentWhat modern science missed in its pursuit of push-and-pull mechanicsHow decision-making, design, and descent all operate on ripple consequences

This is not a conventional book on aerospace. It is a radical rethinking of the **energetic and philosophical foundations** of motion—disguised as rocket science.

Sandeep Chavan's voice is both disruptive and grounded, offering clarity without compromise. His signature UED framework doesn't just add to science—it reshapes it. With previous works redefining time, light, perception, and field dynamics, *Rocket Science: The Ripple Gambling* adds a crucial piece to the puzzle of reality: **how to move without moving.**

Whether you're a physicist, thinker, seeker, or designer of systems, this book will stay in your mind long after the final page. Because in a ripple universe, **you don't land—you realign.**

Are you ready to stop aiming for impact—and start aiming for resonance?

Read more at https://www.sandeepjchavan.com/.

Also by Sandeep Chavan

Chavan's UED research

CMGED Research Report

Chavan's Cosmic Continuum (C^3)

Frequency

Ripples of Reality

Reality Realigned

The Great Realignment

I – The Cosmic Leverage

Chavan's Cosmic Theory of Relativity (CTR)

Chavan's Theory of Light

Feeling Like a Physicist

TIME – The New Definition

$E \neq mc^2$: The Collapse of Classical Assumptions

The Chavan Codex

Echoes of War: The Russia-Ukraine Conflict and Its Global Ripples

Shattered Lives

The War Economy

Borders Redrawn
Arms and Allies
Collateral Damage
Justice in the Ashes

Somewhere in the Darkness of the Cosmos
The Blueprint of the Universe
The Cosmic Breath
From the Edge of the Universe

The Eternal Sage Chronicles
The Sage and the Sovereign
The Family Pillar
The People's Flame
The Warrior's Soul
The Bridging Mentor
The Guiding Light

The Exodus
The Last Cycle: 2075

The Fundamental Theory of Life & Existence
Energy in Action: The Soul Demystified
The Active Mind

Energy in Connection
Death: The Disconnection
Subconsciousness: The Chosen Disconnection
Desires: The Engine of Life

The Restoration of Human Knowledge
Restoring Bhagavad Gita

The War of Thoughts
Cracks in the Known
The Human Game
The Divine Dilemma
The Original Intent
The Answers

Ultra Modern Life Philosophy
The Duality of Laughter and Tears
Take It Easy
Compare to Accept

Standalone
It's Not AI, It's AHI - Amplified Human Intelligence
The Decision Paradox
The Four Sapiens

Malicious Script of Indian Polity
The IIT Legacy & Global Impact
The IIT Legacy & Global Impact
You are just a Version of Your Original
Win the Game You Didn't Choose
Always Keep Your Bags Packed
Reveal to Shield - Learn the Game of Facades
Engage Beyond Elections
Culture, Identity & Change: The Evolution of Indian Society
The Triangular Dynamics
Never Trust Without Doubt
If You Can't Change the Country, Leave the Country
Live Fearless with Fear
India: The Investment Magnet
Realignment
Powering a Billion Dreams
Universal Skill
Extension: From Brush to Algorithm
Human Side of Einstein
The Future Knows No Lines
Let's Bring the Generations to the Table
Unified Energy Dynamics
A Beat That Never Dies
Strategic Horizons
Right From Scratch
The Forgotten Innovators
Physics Fixing
The Savior Illusion
The Selfish Side
The Age of the Weakest Sapiens
Neural Wars

The Absorption: There Is No Death
Strange Therapist
Bubbles: Become a Drop
Beyond the Blue
Half-Scholar
The Law of Truth
Failure
Philosophy 2.0: A Structural Manifesto
Chaos: A Gentle Guide to the Unresolvable
Self-Help: New Syllabus
Rippling
Game: How to Play the Share Market
Out of Syllabus
Saffronized: The Code Behind the Crown
The Anger
Moksha: The Liberation
Chavan vs Einstein: The Universe Was Never Curved
Rocket Science: The Ripple Gambling
The Cosmos: The Still Universe
The Taste of Silence

Watch for more at https://www.sandeepjchavan.com/.

www.ingramcontent.com/pod-product-compliance
Lightning Source LLC
LaVergne TN
LVHW090356210926
848480LV00002B/357